P9-DFT-778

Nuclear Energy

Nuclear Energy

JOHN TABAK, Ph.D.

For Steven Howard Myers. He was a good friend and a very funny man.

NUCLEAR ENERGY

Facts On File, Inc.
An imprint of Infobase Publishing
132 West 31st Street
New York NY 10001

Library of Congress Cataloging-in-Publication Data

Tabak, John.
Nuclear energy / John Tabak.
p. cm.— (Energy and the environment)
Includes bibliographical references and index.
ISBN-13: 978-0-8160-7085-5
ISBN-10: 0-8160-7085-7
1. Nuclear energy. I. Title.
QC792.T33 2009
621.48—dc22 2008026036

Facts On File books are available at special discounts when purchased in bulk quantities for businesses, associations, institutions, or sales promotions. Please call our Special Sales Department in New York at (212) 967-8800 or (800) 322-8755.

You can find Facts On File on the World Wide Web at http://www.factsonfile.com

Text design by Eric Lindstrom
Illustrations by Accurate Art
Photo research by Elizabeth H. Oakes

Printed in the United States of America

Bang Hermitage 10 9 8 7 6 5 4 3 2 1

This book is printed on acid-free paper.

Contents

Preface

Nations around the world already require staggering amounts of energy for use in the transportation, manufacturing, heating and cooling, and electricity sectors, and energy requirements continue to increase as more people adopt more energy-intensive lifestyles. Meeting this ever-growing demand in a way that minimizes environmental disruption is one of the central problems of the 21st century. Proposed solutions are complex and fraught with unintended consequences.

The six-volume Energy and the Environment set is intended to provide an accessible and comprehensive examination of the history, technology, economics, science, and environmental and social implications, including issues of environmental justice, associated with the acquisition of energy and the production of power. Each volume describes one or more sources of energy and the technology needed to convert it to useful working energy. Considerable emphasis is placed on the science on which the technology is based, the limitations of each technology, the environmental implications of its use, questions of availability and cost, and the way that government policies and energy markets interact. All of these issues are essential to understanding energy. Each volume also includes an interview with a prominent person in the field addressed. Interview topics range from the scientific to the highly personal, and reveal additional and sometimes surprising facts and perspectives.

Nuclear Energy discusses the physics and technology of energy production, reactor design, nuclear safety, the relationship between commercial nuclear power and nuclear proliferation, and attempts by the United States to resolve the problem of nuclear waste disposal. It concludes by contrasting the nuclear policies of Germany, the United States, and France. Harold Denton, former director of the Office of Nuclear Reactor Regulation at the U.S. Nuclear Regulatory Commission, is interviewed about the commercial nuclear industry in the United States.

Biofuels describes the main fuels and the methods by which they are produced as well as their uses in the transportation and electricity-production sectors. It also describes the implications of large-scale biofuel use on the environment and on the economy with special consideration given to its effects on the price of food. The small-scale use of biofuels—for example, biofuel use as a form of recycling—are described in some detail, and the volume concludes with a discussion of some of the effects that government policies have had on the development of biofuel markets. This volume contains an interview with economist Dr. Amani Elobeid, a widely respected expert on ethanol, food security, trade policy, and the international sugar markets. She shares her thoughts on ethanol markets and their effects on the price of food.

Coal and Oil describes the history of these sources of energy. The technology of coal and oil—that is, the mining of coal and the drilling for oil as well as the processing of coal and the refining of oil—are discussed in detail, as are the methods by which these primary energy sources are converted into useful working energy. Special attention is given to the environmental effects, both local and global, associated with their use and the relationships that have developed between governments and industries in the coal and oil sectors. The volume contains an interview with Charlene Marshall, member of the West Virginia House of Delegates and vice chair of the Select Committee on Mine Safety, about some of the personal costs of the nation's dependence on coal.

Natural Gas and Hydrogen describes the technology and scale of the infrastructure that have evolved to produce, transport, and consume natural gas. It emphasizes the business of natural gas production and the energy futures markets that have evolved as vehicles for both speculation and risk management. Hydrogen, a fuel that continues to attract a great deal of attention and research, is also described. The book focuses on possible advantages to the adoption of hydrogen as well as the barriers that have so far prevented large-scale fuel-switching. This volume contains an interview with Dr. Ray Boswell of the U.S. Department of Energy's National Energy Technology Laboratory about his work in identifying and characterizing methane hydrate reserves, certainly one of the most promising fields of energy research today.

Wind and Water describes conventional hydropower, now-conventional wind power, and newer technologies (with less certain futures) that are being introduced to harness the power of ocean currents, ocean waves, and the temperature difference between the upper and lower layers of the ocean. The strengths and limitations of each technology are discussed at some length, as are mathematical models that describe the maximum amount of energy that can be harnessed by such devices. This volume contains an interview with Dr. Stan Bull, former associate director for science and technology at the National Renewable Energy Laboratory, in which he shares his views about how scientific research is (or should be) managed, nurtured, and evaluated.

Solar and Geothermal Energy describes two of the least objectionable means by which electricity is generated today. In addition to describing the nature of solar and geothermal energy and the processes by which these sources of energy can be harnessed, it details how they are used in practice to supply electricity to the power markets. In particular, the reader is introduced to the difference between base load and peak power and some of the practical differences between harnessing an intermittent energy source (solar) and a source that can work virtually continuously (geothermal). Each section also contains a discussion of some of the ways that governmental policies have been used to encourage the growth of these sectors of the energy markets. The interview in this volume is with John Farison, director of Process Engineering for Calpine Corporation at the Geysers Geothermal Field, one of the world's largest and most productive geothermal facilities, about some of the challenges of running and maintaining output at the facility.

Energy and the Environment is an accessible and comprehensive introduction to the science, economics, technology, and environmental and societal consequences of large-scale energy production and consumption. Photographs, graphs, and line art accompany the text. While each volume stands alone, the set can also be used as a reference work in a multidisciplinary science curriculum.

Acknowledgments

The author is particularly grateful for the help of the following individuals: Harold Denton, former director of the Office of Nuclear Reactor Regulation, who was extremely generous with his time and knowledge; Ken McDonnell, senior media relations specialist at ISO New England, whose research and thoughtful replies were extremely valuable; Elizabeth Oakes, who researched the photos used in this volume; Leo Christian-Tabak, for help with the artwork; and Frank Darmstadt, executive editor, for the vote of confidence.

Introduction

After a long period of little to no growth, the commercial nuclear power industry seems ready to begin another period of robust growth. New reactors have been built or are currently under construction in Japan, Taiwan, China, Finland, and France, for example, and new reactors are under consideration in the United States, Great Britain, and Canada. This "nuclear Renaissance" has generated considerable interest.

Commercial nuclear power has long been one of the most controversial of all power-generation technologies, but often the debate has not been a particularly informed one. Understanding the contribution that nuclear power makes today and the potential of nuclear power in the future requires an appreciation of various branches of physics, engineering, mathematics, economics, the environment, and the way that power is supplied to the grid. It also helps to apply the same ideas and standards to the evaluation of competing technologies. Neither an endorsement nor a condemnation of nuclear power, *Nuclear Energy,* one book in this multivolume set, aims to provide as much objective information about the subject as space allows. It is unique in that it is both accessible—it assumes only a modest knowledge of high-school algebra on the part of the reader—and very broad in scope.

Chapter 1 is an overview of the theory by which *thermal energy* is converted into electrical energy, because nuclear reactors are, first and foremost, *heat engines* and have much in common with the many other types of power plants such as natural gas, geothermal, coal, oil, and concentrating solar-power systems. Chapter 4 describes five different types of nuclear power plants to show how this is accomplished in practice.

What distinguishes nuclear reactors from other types of power plants that convert thermal energy into electrical energy is, of course, the source of their heat. Nuclear power plants produce heat by releasing energy from the nuclei of atoms through a process called atomic *fission.* The process, which releases enormous amounts of energy per unit mass of fuel, has numerous consequences for the environment, for the consumer, and, potentially, for national security. For this reason, a great deal of attention is paid to the fission process in chapters 2 and 3.

Safety must be the most important consideration in the design and operation of nuclear power plants because they produce so much thermal energy. Some of the mechanisms by which safe operation is achieved are described in chapter 5. Because the fission process produces highly toxic by-products, the wastes produced during the fission process must be handled with great care. Some governments have chosen to recycle used reactor fuel because when the fuel is removed from the reactor most of the energy in the fuel has not been utilized. Other nations have decided as a matter of policy to simply bury the used fuel. Both strategies have important implications for the environment and are discussed in chapter 6.

Nuclear power is just one of several methods for producing electricity and in this sense nuclear power plants must compete with other power-generation technologies, all of which can be compared in terms of reliability, cost, and effects on the environment. Although commercial nuclear power plants are currently some of the most reliable, efficient, and least expensive plants to operate in the United States, no new plants have been ordered in decades. Some of the reasons for this are examined in chapter 7.

Finally, national policies have a tremendous impact on the development of the power markets, and no developed nation has shown itself willing to leave the development of its energy industries to "the market." Chapter 8 describes and contrasts the nuclear energy policies of Germany, the United States, and France. The book closes with an interview with Harold Denton, the former director of the Office of Nuclear Reactor Regulation at the U.S. Nuclear Regulatory Commission, about the commercial nuclear industry.

Billions of people have come to rely on adequate, affordable, and reliable supplies of electricity, and billions more want access to those supplies. Everyone understands what access to electricity means for the quality of life. Meeting this demand in a way that minimizes the effects of electricity production on the environment is one of the central problems of this century. It is the responsibility of every citizen to learn about the issues, the technologies, and consequences of power generation and to contribute to the discussion about the best ways of producing the energy on which the world depends.

A Prehistory of Nuclear Power

A nuclear power plant is a single-use machine. Its sole purpose is to convert thermal energy into electrical energy, and in this sense nuclear reactors have much in common with other commercial power plants. Coal-fired, natural gas–fired, oil-fired, geothermal, and even some solar-powered electricity-generating stations operate according to the same general three-step procedure: (1) Thermal energy is produced, (2) the thermal energy is converted into the energy of motion, and (3) the energy of motion is converted into electrical energy. What distinguishes nuclear plants from all other power-generation technologies is the process by which they generate heat. Nuclear plants depend upon the phenomenon of nuclear fission rather than, for example, combustion, which is a chemical reaction. Understanding the consequences of the use of nuclear fission to generate heat is crucial to understanding the nature of nuclear power. But in order to place nuclear power in context relative to other power-generation technologies, it is just as important to understand what characteristics nuclear plants share with more conventional power plants, and that is one of two topics addressed in this chapter; the other topic shows how the idea of using nuclear energy to produce electricity first arose.

THE FIRST HEAT ENGINES

Steam engine technology is about three centuries old. Early engines were extremely crude, of course, but because they demonstrate in a very transparent way some of the concepts that are most important in understanding the workings of nuclear power plants, they

Newcomen steam engine—notice the size of the pump compared to the man *(Public domain)*

are well worth examining. The English inventor Thomas Newcomen (1663–1729) built one of the earliest steam engines. In the accompanying cutaway diagram notice that Newcomen's steam engine, which is located on the right side of the central pillar, is attached by a chain to one side of a massive rocker arm. The left side of the rocker arm is attached—again by a chain—to a water pump, which is located on the left side of the drawing. The rocker arm, which works like a pump handle, connects the steam engine to the water pump and is weighted so that when released it tips toward the pump side. The steam engine moves the rocker arm, and the up-and-down motion of the rocker arm drives the pump. Although the concept may seem simple, the application could not have been more important for the development of British industry: By the time Newcomen invented his engine, the British had already mined the deposits of coal that were close to the surface of the ground. They needed to dig deeper. The problem that they faced was that the deeper mines tended to fill with water. This is the reason that early steam engines were so important. They made deep mining possible and, in fact, all early steam engines were used to pump water out of coal mines. In this illustration the pump is connected to the water in the mine by a pipe. In the diagram, the valve labeled *H* regulates the flow through that pipe. The engine worked in the following way:

1. Coal is burned inside the boiler to generate heat.
2. Water is pumped inside the boiler, where it turns to steam.
3. As the rocker tips toward the pump side—as previously mentioned, it is weighted so that it tips toward the pump in the absence of other forces—the piston in the steam cylinder begins to move upward, and the valves operate in the following sequence:
   (i) Valve *A* opens—and valves *B* and *C* close—so that steam rushes from the boiler into the cylinder, filling the cylinder with steam.

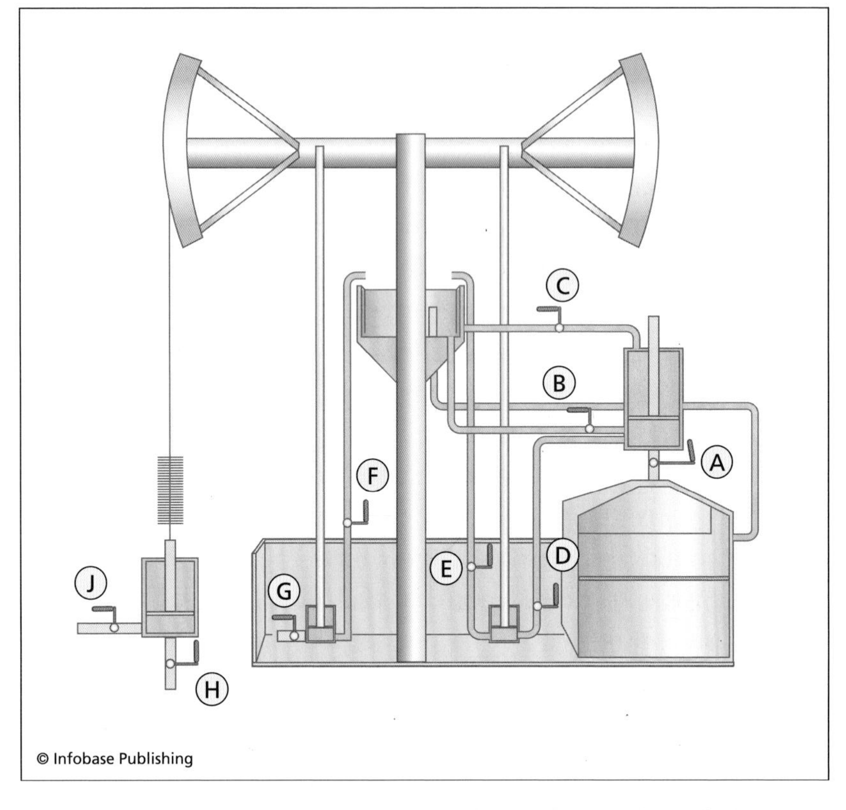

Cutaway diagram of Newcomen's engine. This modern-looking idealized version of Newcomen's engine illustrates the method by which Newcomen converted thermal energy into work.

(ii) Valve *J* opens—and valve *H* closes—so that as the heavy pump piston descends, water is forced out of the pump cylinder through valve *J*. (This is the water that was previously in the mine. It was drawn up out of the mine on a previous stroke of the piston.)
(iii) Valve *D* opens—and valve *E* closes—so that any water present in the steam cylinder due to the condensation of steam is drawn out of the steam cylinder.
(iv) Valve *F* opens—and valve *G* closes—so that the auxiliary pump moves water into the reservoir located at the top of the pillar.

4. When the piston reaches its maximum height inside the steam cylinder—that is, the left side of the rocker arm is as low as it can go and the right side of the rocker arm is as high as it can go, the steam cylinder has filled with steam. Now

   (i) Valve *A* closes, shutting off the steam from the boiler, and valves *B* and *C* open so that water can flow from the water reservoir into and onto the steam cylinder. The water is released in the form of a spray. This cools the cylinder and the steam inside it. As a consequence, the steam quickly condenses into liquid water. As the steam condenses, the pressure inside the cylinder drops below atmospheric pressure. The atmosphere pushes down on the top of the cylinder with sufficient force to drive the piston downward, pulling the right side of the rocker arm down.

   (ii) As the piston descends, valve *D* closes, and valve *E* opens, allowing water to flow from the condensate pump, located on the right side of the drawing, up into the water reservoir.

   (iii) At the auxiliary pump, located on the left side of the drawing, valve *F* closes and valve *G* opens. As the piston in the auxiliary pump rises, water is drawn into the auxiliary pump. It will be pumped into the water reservoir on the next stroke of the engine.

   (iv) Valve *J* closes, and valve *H* opens, and water is drawn upward from the mine and into the cylinder to be expelled outward on the next stroke.

5. When the piston in the steam cylinder reaches its lowest point, the cycle repeats.

Newcomen's engine is a *conversion* device. It converts thermal energy into the up-and-down motion of the rocker arm. This conversion requires a "working fluid," in this case water, to transport thermal energy from the boiler to the steam cylinder. It is through the addition and subtraction of thermal energy to the working fluid that work is performed: Thermal energy is transferred to the water at the boiler; the additional energy causes the liquid water to change to steam, and the steam flows into the cylinder. Much of the thermal energy carried by the steam is transferred to the cool water that is sprayed inside the cylinder and onto its exterior, causing the steam to change back into liquid water; the pressure drops; the atmosphere pushes down on the cylinder, and the pump raises water from the mine. The complicated sequence of valve, lever, and piston motions was Newcomen's way of accomplishing the simply stated task of pumping water.

Nineteenth-century engineers discovered that other working fluids could be used in place of water, but many large heat engines, even today, continue to use water because it is both plentiful and cheap. Newcomen's engine even recycles the condensed steam: Water flows from the water reservoir to the steam cylinder through the condensate pump and then back into the water reservoir. (Because of losses due, for example, to evaporation and leaks in the system, some additional water, supplied by the auxiliary pump, had to be delivered to the reservoir to keep the engine operating.)

It is important to recognize that not all of the thermal energy was converted into the motion of the rocker arm. Some thermal energy escaped from the boiler into the surrounding air; some escaped through the sides of the steam pipes and some through the sides of the steam cylinder. And the water that was sprayed inside the cylinder to cause the steam to condense also cooled the sides of the cylinder and the piston. The cooled steam cylinder and piston also represent a loss because some of the thermal energy from the "new" steam—the steam that would enter the cylinder during the next cycle—flowed out of the steam and into the cooled cylinder and piston. (Thermal energy always flows from warmer regions to cooler ones.) Consequently, the temperature of the steam dropped and there was some premature condensation. This, too, is a waste in the sense that premature condensation does not help draw the steam piston down from its maximum height. The successful operation of any steam engine requires the careful management of the thermal energy carried by the steam.

Wasted thermal energy—wasted in the sense that it does not contribute to the movement of the piston inside the steam cylinder—raises the concept of steam engine *efficiency.* For the Newcomen engine, how much of the thermal energy in the boiler was actually used to raise water from the mine? The answer to this question is a measure of the efficiency of the engine. Because the operators had to pay for all of the fuel that they burned in the boiler, they wanted to convert as much of the thermal energy as possible into the motion of the rocker arm. Wasted thermal energy meant wasted money. Although neither Newcomen nor any of his contemporaries could calculate the efficiency of the engine—they were still unclear about the relationship between thermal energy and the work performed by the engine—they were aware that the Newcomen engine consumed a great deal of fuel for the amount of water that it lifted. Engine inefficiencies were reflected in high fuel costs. It has since been calculated that only about 1 percent of the thermal energy generated in the boiler was converted into work. The rest, about 99 percent, was wasted.

Finally, notice that Newcomen's engine only requires that the boiler be hot enough to generate the necessary steam. One can generate this heat by burning coal or wood or peat. Although engineers of this period lacked both the scientific insight and the technology to do so, there is no theoretical reason why the thermal energy that the Newcomen engine required could not have been generated via nuclear fission. All steam engines require a heat source, but provided the fuel can generate enough thermal energy fast enough, any heat source will do.

Despite its shortcomings, the Newcomen steam engine was a tremendous accomplishment at the time of its invention and its design was unsurpassed for more than 50 years. Steam engines were largely redesigned and greatly improved by the Scottish inventor James Watt (1736–1819). Some of his innovations can be found in a much more sophisticated form in most nuclear power plants.

One of Watt's most important innovations was to add a chamber that was physically separate from the steam cylinder but connected to it via a tube. Access to this cool chamber, called a condenser, was controlled by a valve. By creating a partial vacuum inside the condenser, Watt caused steam to flow from the steam cylinder to the condenser as soon as the valve connecting them was opened. Upon reaching the condenser, the steam would condense and the pressure in the steam cylinder would drop rapidly. By condensing the steam inside the condenser and away from the steam cylinder, Watt maintained the steam cylinder at a consistently high temperature. This innovation greatly increased the efficiency of Watt's engines: They converted between 2 and 3 percent of the thermal energy generated by the boiler into work. They also used only about one-third of the fuel of the Newcomen engine, a tremendous cost savings for the user, and, in fact, Watt's fees for his engines were calculated in part according to the savings in fuel enjoyed by the user.

Second, notice that Newcomen's engine could only pull *down* on the rocker arm; that is, the power stroke was the downward stroke of the piston. This is apparent from the period drawing at the beginning of the chapter: The steam piston is connected to the rocker arm by a chain, and one cannot push with a chain. By contrast, Watt sealed the entire piston inside the cylinder and devised a way to inject steam alternately above and below the piston. As a consequence, Watt's engines could push as well as pull on the rocker arm.

Third, Watt patented a device called a sun and planet gear system. As shown in the accompanying illustration, this system enabled him to change the linear, or straight line, motion of the piston into a rotary motion, a tremendous advantage over Newcomen's simple up-and-down motion because a turning shaft can be used to

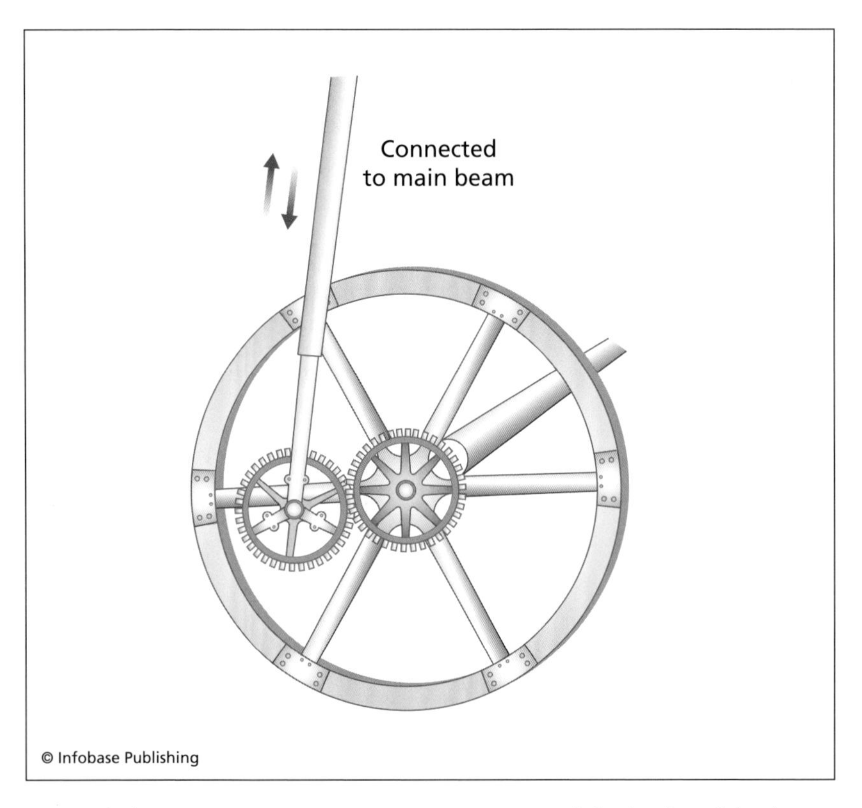

Sun and planet gear system: Watt's invention turned the back-and-forth motion of the piston into rotary motion

drive almost any piece of machinery. Initially, this innovation was applied to milling flour, but it soon found applications in a wide variety of industrial situations.

Watt, of course, did not improve every aspect of engine performance. All of Watt's steam engines operated on the difference in pressure between the interior of the steam cylinder and the environment—that is, the difference between the partial vacuum created inside the cylinder and atmospheric pressure. Because the outside air pressure is fixed—in the sense that it cannot be changed by the engine operator—and the interior of the steam cylinder cannot have a pressure less than zero, the largest pressure difference these engines can generate between the interior and exterior of the steam cylinder is equal to one atmosphere. Watt knew that one could manufacture a more powerful engine by using high-pressure steam—even during his lifetime other engineers designed and built high-pressure steam engines—but for safety reasons Watt remained committed to building engines that operated at atmospheric pressure. High pressures were dangerous, he knew, because the materials and manufacturing techniques in use at the time meant that sometimes a high-pressure boiler would explode, injuring or killing the operator.

But from a design point of view, high pressures are highly desirable because even with a small cylinder, a high-pressure steam engine can exert a great deal of force. And smaller, more portable steam engines opened up new applications: They could be used to drive trains and ships, for example, so despite concerns about safety, high-pressure engines were soon manufactured. Often high-pressure engines were made with multiple cylinders. The cylinders worked in series: High-pressure steam was pumped inside the first cylinder, in which some—but only some—of the thermal energy was converted into the up-and-down motion of the first cylinder. When the first cylinder had completed its job, a valve would open and somewhat cooler steam flowed from the first cylinder to the second cylinder where it drove a second piston. The second piston was designed to work on the lower-pressure, lower-temperature steam that it obtained from the first piston. (The steam was cooler because some of the thermal energy it originally carried had been converted into work in the first cylinder.) Sometimes the second cylinder was connected to a third and occasionally even a fourth. Each successive piston was designed to use steam at successively lower temperatures and pressures, which it received from the cylinder immediately preceding it in the sequence. The condenser was connected to the last and lowest-pressure cylinder. These technological innovations enabled engineers to create more powerful, more efficient engines that made better use of the thermal energy produced by the boiler. The result was a revolution in the industrial and transportation sectors of Britain and later the entire world.

CARNOT AND THE EFFICIENCY OF HEAT ENGINES

The industrial and military advantages that the British enjoyed over their competitors during the late 18th and early 19th centuries resulted, at least in part, from their comparative mastery of steam engine technology, and this fact was not lost on the young French engineer Nicolas-Léonard-Sadi Carnot (1796–1832). Possessed of a creative mind and motivated by a sense of patriotism, Carnot set out to discover the scientific principles that governed heat engines. (Carnot made a distinction between steam engines, which use steam to transfer thermal energy from the boiler to the steam cylinder, and heat engines, a more general term for any device that converts thermal energy into motion.) The result was his book, *Reflections on the Motive Power of Fire*. First published in 1824, it is a short and profound scientific treatise on the theory of heat engines. The book is still in print—it currently sells more copies each year than it ever sold during Carnot's lifetime, when it was largely ignored—and Carnot's work is now recognized as containing the first essentially accurate description of the fundamental properties of heat engines. It is as relevant to nuclear reactors as it is to Newcomen's engine.

Heat engines are a particularly valuable technology, Carnot writes, because in contrast to windmills or waterwheels, which depend on the wind or the level of water in streams, factors over which humans have limited control, heat engines can be made to work whenever and wherever the user needs power, provided, of course, that fuel is available. Heat engines are, as a consequence, extremely important power sources because they allow the user to match power supply with power demand in a way that other technologies of the day could not.

Because his ideas were new, the vocabulary needed to describe his insights had not yet developed. Carnot's explanations are, therefore, often expressed as analogies between heat and water and between heat engines and waterwheels, a technology with which his readers were familiar. A waterwheel, for example, will not work in a motionless pool of water. Waterwheels work *between* levels of water. As water falls (or flows) from a higher to a lower level, some of that energy of motion can be used to turn the wheel. In other words, waterwheels convert some of the energy of falling water into rotary motion. Similarly, heat engines cannot operate at one temperature no matter how high or low that temperature is. Heat engines require a temperature difference in order to function. Heat always flows from a warmer region to a cooler one, just as water always flows downhill, and by placing a heat engine *between* the warmer and cooler regions, some of the flowing heat can be converted into the energy of motion. (Carnot conceived of heat as an invisible fluid that all bodies possessed. As heat, which he called "caloric," flowed into a body, the body's temperature increased, and it expanded, he believed, to accommodate the additional caloric. Similarly, when caloric flowed out of a body, it contracted as it cooled. Curiously, Carnot's ideas about caloric have since been disproved, but his ideas about the nature of heat engines, which were based in part on his understanding of caloric, have stood the test of time.)

The efficiency of a waterwheel—that is, the amount of work that it performs for each unit of water that passes through it—was limited, Carnot knew, by the difference in height between the upper and lower reservoirs of water that it connected. If the water fell a great distance, it could exert more force on the wheel. One unit of water could, therefore, perform a great deal of work provided it fell from a great enough height. If, on the other hand, the upper and lower reservoirs of water were at almost the same height, then water flowed gently past the wheel, and many units of water would be required in order that the wheel perform the same amount of work as was performed by one unit of water falling a great distance. Carnot asserted that the same principle applied to heat. The efficiency of a heat engine is limited by what he called the "fall" in temperature between the high-temperature reservoir and the low-temperature reservoir. The larger the temperature difference, the more work could be extracted from each unit of heat.

The ratio formed by the amount of work done by an engine to the amount of heat provided to an engine, which is called the "heat absorbed," is a measure of the efficiency of the engine:

$$Efficiency = \frac{Work\ Done}{Heat\ Absorbed} \quad \textbf{(1.1)}$$

One consequence of Carnot's theory is that the maximum efficiency of any heat engine is determined by this simple equation:

$$Maximum\ Efficiency = 1 - \frac{T_C}{T_H} \quad \textbf{(1.2)}$$

where T_C is the temperature of the cooler reservoir, usually taken to be the temperature of the environment, and T_H is the temperature of the hotter reservoir, usually taken to be the temperature of the boiler. (Both temperatures are measured in degrees kelvin, a temperature scale in which a change of one degree kelvin equals a change of one degree Celsius and where 0 degrees Celsius equals 273.15 degrees kelvin.) Notice that when T_C equals T_H, the efficiency of the engine is zero, and as T_H increases relative to T_C so does the maximum efficiency of the engine. There is, however, no operating temperature T_H that will make any heat engine 100 percent efficient. Consequently, some heat is always wasted. It is also important to emphasize that equation (1.2) only establishes a limit on the maximum efficiency of a heat engine. While Carnot proved that no engine can be more efficient than what equation (1.2) predicts, in practice many heat engines are much less than maximally efficient.

Finally, it is important to emphasize that in the preceding discussion no reference is made to the type of fuel that the heat engine uses. In fact, the type of fuel one uses to generate the thermal energy needed to run the engine is completely irrelevant when determining the maximal efficiency of the engine.

Steam engine technology continued to evolve throughout the 19th century, and the favored fuel for these machines was coal because about 65 percent more thermal energy can be generated from burning a ton of coal than a ton of wood. (A more exact figure depends upon the type and moisture content of the wood and coal supplies under consideration.) Engineers sometimes describe this situation by saying that coal has a higher *energy density* than wood. Late in the 19th century, however, scientists discovered another class of materials with energy densities that were far greater than those of coal.

RADIOACTIVE MATERIALS AS SOURCES OF THERMAL ENERGY

Radioactive decay involves the spontaneous disintegration of an atomic *nucleus.* As the nucleus disintegrates, various particles as well as high-energy electromagnetic waves are released. *Radioactivity* is not a new phenomenon. Radioactive materials have always been present everywhere in the environment—in the atmosphere, on land, and in the oceans—but until the latter years of the 19th century the phenomenon of radioactivity had escaped scientific attention because it usually exists at such low levels that its effects are difficult to observe.

As a subject of scientific interest, radioactivity came to the fore when the French scientist Antoine-Henri Becquerel (1852–1908) performed a series of experiments whose purpose was to investigate X-rays. In these experiments he wrapped unexposed photographic plates in opaque paper so that no light could reach them and placed certain materials near the wrapped plates. He discovered that when he placed compounds containing the metal uranium near the plates, they emitted rays that passed through the paper and exposed the plates. He found that he could take primitive pictures with these

Antoine-Henri Becquerel, ca. 1904 *(Library of Congress Prints and Photographs Division)*

rays: By placing a coin between the material emitting the rays and the photographic plate, for example, he found that the coin blocked the rays. The result was that the plate was exposed everywhere except for the area beneath the coin. Although he misinterpreted the nature of the phenomenon he had discovered, his results attracted the attention of the most successful husband-and-wife team in the history of science, the French scientist Pierre Curie (1859–1906) and the Polish-born scientist Marie Curie (1867–1934). (Marie Curie later coined the term *radioactivity*.)

The phenomenon observed by Becquerel was a weak one because natural uranium emits few radioactive particles per unit time. Scientists describe this situation by saying that natural uranium has a low specific *activity*. Intrigued by Becquerel's experiments, Marie Curie began to investigate the mineral pitchblende, which

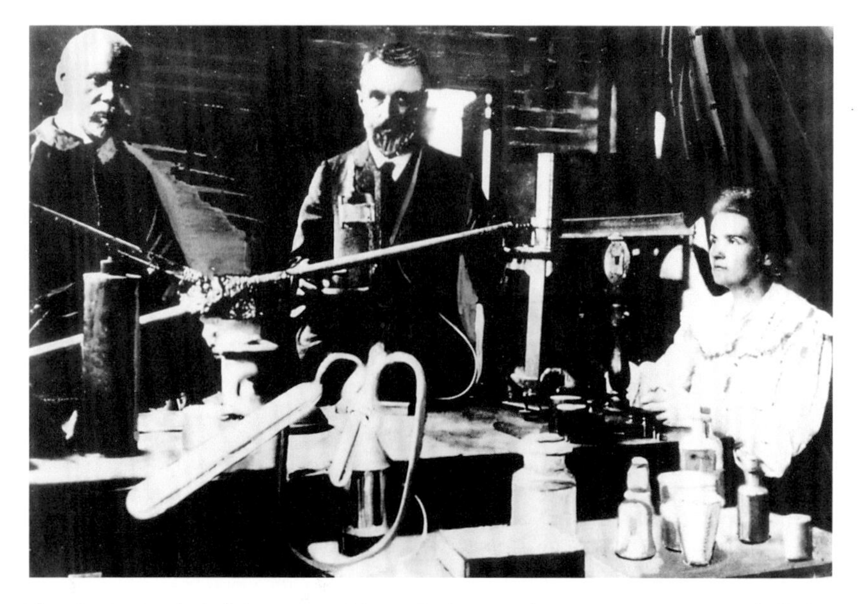

The Curies in their laboratory. Marie Curie was the most important of all the early researchers in the field of radioactive phenomena. The man on the left is unidentified. *(AIP Emilio Segrè Visual Archives)*

has a higher specific activity than uranium. While Pierre Curie concentrated on identifying the physical properties of radiation, Marie sought to isolate the elements responsible for emitting the higher levels of radiation. Her work culminated in the discovery of two new elements, polonium and radium. The element radium was of particular interest because compared to uranium, radium has a very high specific activity; in other words, a unit mass of radium emits a great deal more radioactivity per unit time than does a unit mass of natural uranium.

Physicists of the time were astonished by what Marie Curie had discovered. Not only did the radium continuously emit relatively large amounts of energy, but it emitted the energy for long periods of time without changing in any perceptible way. Keep in mind that the principle of conservation of energy, which states that energy can neither be created nor destroyed, meant that energy was, in effect, draining away from her sample of radium. As Curie knew, this continual emission of energy must eventually leave a perceptible change in the physical properties of her radium sample. The fact that she could not quickly detect any change in the radium meant that the amount of energy contained in a small piece of radium must be very large. (By way of analogy, imagine trying to drain a lake with a garden hose. One can pump water through the hose for a very long time without noticing any change in the level of the lake because the capacity of the lake is large compared to the rate of flow through the hose. If one switched to a very large hose and still could not detect a change in lake level after a long period of pumping, one could be sure, even without knowing the capacity of the lake, that the lake contained a great deal of water.) In 1904, Marie Curie expressed her insights in these words:

> Radium possesses the remarkable property of liberating heat spontaneously and continuously. A solid salt of radium develops a quantity of heat such that for each gram of radium contained in the salt there is an emission of one hundred calories per hour. Expressed differently, radium can melt in an hour its weight in ice. When we reflect that radium acts in this manner continuously, we are amazed at the amount of heat produced, for it can be explained by no known chemical reaction. The radium remains apparently unchanged. If, then, we assume that it undergoes a transformation, we must therefore conclude that the change is extremely slow, in an hour it is impossible to detect a change by any known methods.

The ability of radium to emit large quantities of energy for prolonged times without perceptible change led the British physicist Ernest Rutherford (1871–1937) to speculate on the possibility of harnessing radioactive materials for warfare and for work. If one could release all of the energy at once, he wrote, one could create a tremendous bomb. This, after all, is what a bomb does: It releases a great deal of energy over a very short period of time. If, on the other hand, one could increase the rate of energy release in a gradual and

Global Warming and Nuclear Power

The idea behind global warming is simple enough. Earth's atmosphere retains some of the heat that Earth receives from the Sun. The amount of heat retained depends upon the chemical composition of the atmosphere. If the chemical properties of Earth's atmosphere are changed, then the thermal properties of the atmosphere—that is, its heat-retention properties—change as well.

Burning fossil fuels releases carbon dioxide into the atmosphere. The combustion reaction in which fossil fuels are consumed produces, in the main, two products: carbon dioxide and water. At ordinary temperatures and pressures, carbon dioxide is a gas, and so burning fossil fuels releases carbon dioxide gas into the air. There is ample evidence that carbon dioxide levels in the atmosphere have increased as a result. Consequently, the thermal properties of the atmosphere have changed. More carbon dioxide in the atmosphere means more of the energy that Earth receives from the Sun is retained. Temperatures are, as a consequence, rising.

Many questions about the effects of a warmer atmosphere on storm systems, ocean currents, and rainfall patterns have yet to be resolved. But there will be effects—there have, in fact, already been numerous easy-to-measure effects—and over time these effects will increase in magnitude as atmospheric carbon dioxide levels continue to increase. In fact, the rate of fossil fuel consumption continues to increase as more of the world's population becomes accustomed to lifestyles previously identified only with developed nations. The reason? Western lifestyles are heavily dependent on the consumption of enormous quantities of fossil fuels.

Nuclear power plants produce no carbon dioxide. They depend, instead, on an entirely different mechanism for producing heat than the

controlled way, one would have a source of power sufficient to supply the largest of cities with its power needs. If only one could control the rate at which radium emitted its energy, Rutherford speculated, a small mass of radium could replace a very large mass of coal.

chemical processes associated with burning fossil fuels. Nuclear plants are, therefore, capable of producing large amounts of electricity without affecting the global climate. By contrast, replacing an average size nuclear plant with coal-burning power stations would entail burning 2.8 million tons (2.5 million metric tons) of coal each year. Nuclear power plants are, therefore, one possible answer to the questions posed by Raymond L. Orbach, Under Secretary for Science at the U.S. Department of Energy, in a 2007 speech at Iowa State University: "We must find a way to meet the increasing demand for energy without adding catastrophically to atmospheric carbon dioxide. The world, therefore, has a twofold problem: Where will this new energy come from, and how can it be carbon-free?"

To see the impact of nuclear plants on carbon emissions, consider what has happened in France: Between 1970 and 1995 France's population increased by 13 percent; the size of its economy increased by 71 percent; the amount of electricity it produced increased by 214 percent, and its carbon dioxide emissions *decreased* by 16 percent. This happened because during this same period of time the percentage of electricity generated by nuclear power in France rose from 6 percent to 77 percent. Nuclear output has since held steady at about 80 percent of France's total electrical output. The remaining 20 percent is evenly divided between hydroelectric and fossil fuels. Despite much debate about the importance of reducing greenhouse gas emissions among the major developed nations, only France has managed to make substantial reductions in emissions, because only France has almost eliminated its dependence on fossil fuels for electricity generation.

In principle, Rutherford was right. Nor was he alone. The British writer H. G. Wells (1866–1946) had the same sorts of insights. But while their instincts were correct, Rutherford and his contemporaries were wrong about the power source. It would not be radium that would power cities and serve as fuel for a new and more powerful type of bomb but rather the radioactive metals uranium and the yet-to-be-created plutonium.

To understand how the ideas of this chapter come together, consider the simplified schematic of one of the most common types of nuclear reactor, a boiling water reactor. (See the accompanying illustration.) Boiling water reactors were first pioneered in the United States, where a number of them continue to produce electricity. Japan has recently installed several boiling water reactors of a more advanced design. Both the older and newer reactors demonstrate all of the ideas described in this chapter. But while many of the design concepts employed in these reactors are the same as what one finds in the early steam engines, the technology employed in nuclear reactors has, not surprisingly, been greatly modified to make better use of the tremendous steam temperatures and pressures that these plants are capable of producing.

Apart from the fuel used by these reactors and the scale on which they operate, the most obvious difference between Newcomen's and Watt's engines and contemporary boiling water reactors is the *turbine.* The turbine is the analogue of Watt's sun and planet gear system. It converts the linear or straight-line motion of flowing steam into rotary motion. Turbines are designed to rotate continuously in one direction and at a steady speed and are driven by the high-pressure steam that flows in a continual torrent out of the *pressure vessel,* the analogue of Newcomen's boiler. Designers prefer to use turbines because they have found that the back-and-forth motion of pistons, the continual opening and closing of valves, and the sun and planet gear systems are poorly suited for electrical power generation, a process in which all components must operate for pro-

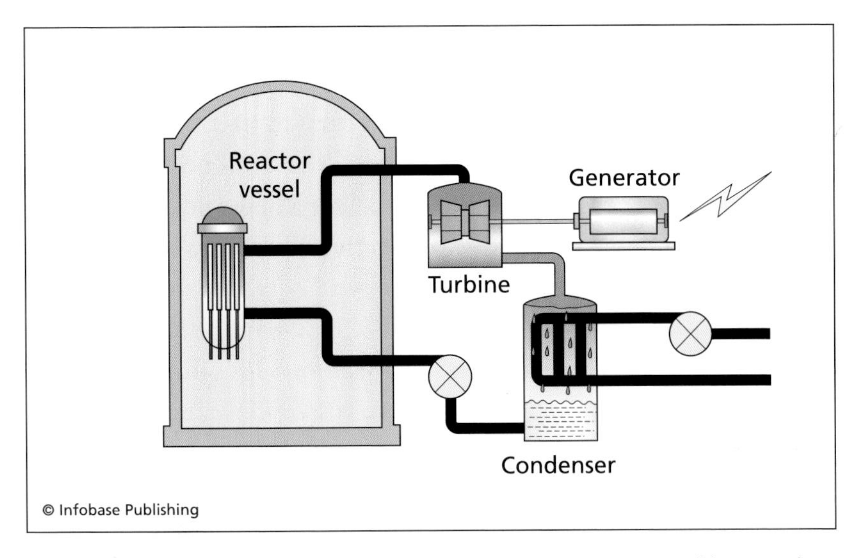

A boiling water reactor is, in the end, a type of steam engine, and has much in common with other steam engines.

longed periods of time under conditions of high temperature, high pressure, and other large forces.

For boiling water reactors, the cycle by which electricity is produced works as follows:

1. Heat is generated by the nuclear fuel, which is located in the reactor vessel.
2. Water is pumped into the reactor vessel, where some of it is converted to high-temperature, high-pressure steam.
3. The steam, which leaves the pressure vessel at about 550°F (290°C) and at a pressure of about 71 atmospheres, travels along the steam line until it reaches the turbine.
4. Upon passing through a valve, the expanding steam pushes against the blades of the turbine, causing the turbine to spin. The turbine is connected to the electrical generator, the device that actually produces electricity,

by a shaft. (The entire purpose of any power plant is to drive the generator.)

5. After exiting from the turbine, the steam enters the condenser, where it is cooled and in the process condenses back to the liquid state. Cooling occurs when the steam comes into contact with a device called a heat exchanger. The heat exchanger is kept cool by water that circulates through it continuously. The water that cools the heat exchanger is physically separate from the water that passes through the pressure vessel. Only heat, not water, is exchanged inside the condenser. One purpose of the condenser is to reduce the back pressure of the steam after it has exited the turbine. In a way that is conceptually similar to the function of Watt's condenser, the condenser in the boiling water reactor enables the turbine to make maximum use of the steam.
6. From the condenser the liquid water is pumped back into the pressure vessel, and the cycle repeats.

Boiling water nuclear power plants have a great deal in common with their Age of Steam predecessors. They are machines that use thermal energy to create steam to do mechanical work. The characteristic that most distinguishes these nuclear power plants from other types of heat engines is the fuel that they use to generate thermal energy, but the consequences of using nuclear fuel are profound.

The Physics of Nuclear Fission

To appreciate the way in which nuclear reactors produce their power, it is necessary to understand something about the processes of fission, *transmutation,* and the conditions under which these processes occur. That is the goal of this chapter.

Atoms are the fundamental units out of which all elements are composed. There are, roughly speaking, as many different kinds of atoms as there are elements—114 kinds at the time this is written. (Some elements, because they do not occur naturally on Earth, must be created in order to be observed, and some of these artificially created elements exist for only a tiny fraction of a second before they spontaneously self-destruct.)

ATOMIC PARTICLES

For purposes of this section, an atom may be imagined as a sort of mini–solar system. This solar system model of atomic structure is

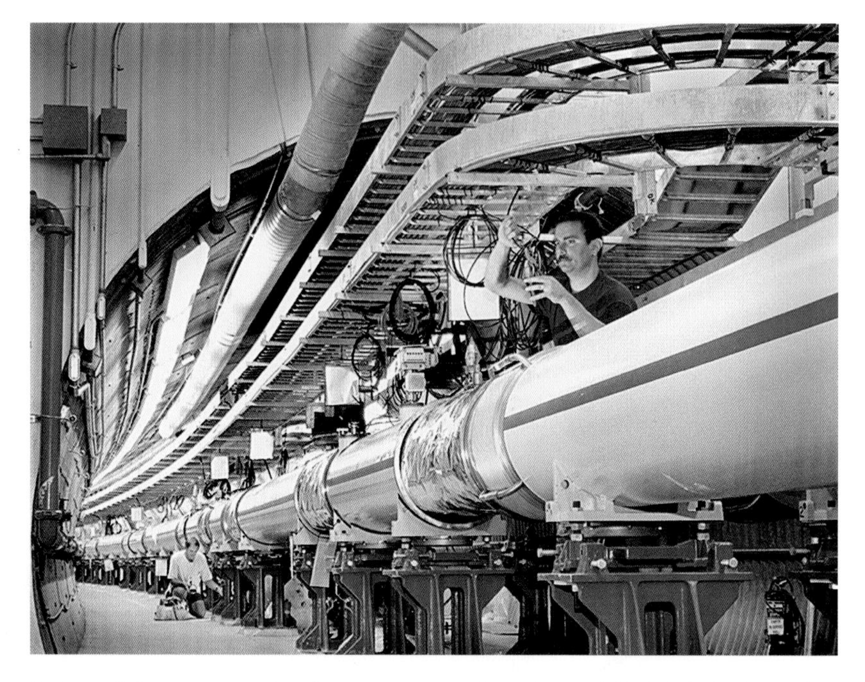

A view of the superconducting magnets at Brookhaven's Relativistic Heavy Ion Collider. These large machines are used to obtain insight into atomic structure. *(Brookhaven National Laboratory)*

called the Bohr model after its inventor, Niels Bohr (1885–1962), a Danish physicist and one of the most successful of the early pioneers in the field of atomic physics. In the Bohr model, a relatively large, dense mass called the nucleus lies at the center of each atom. Particles called *electrons,* each of which is very small compared to the nucleus, orbit the central mass of the atom at certain specific distances. In the Bohr model the nucleus corresponds to our Sun, and the electrons correspond to planets moving about the Sun. As is true for our solar system, where most of the mass is located within the Sun at the center of the system, most of the mass of the atom lies within the nucleus, and as with the solar system every atom consists largely of empty space.

The nucleus of every atom is composed of two types of particles, *neutrons* and *protons.* All atomic nuclei contain at least one proton,

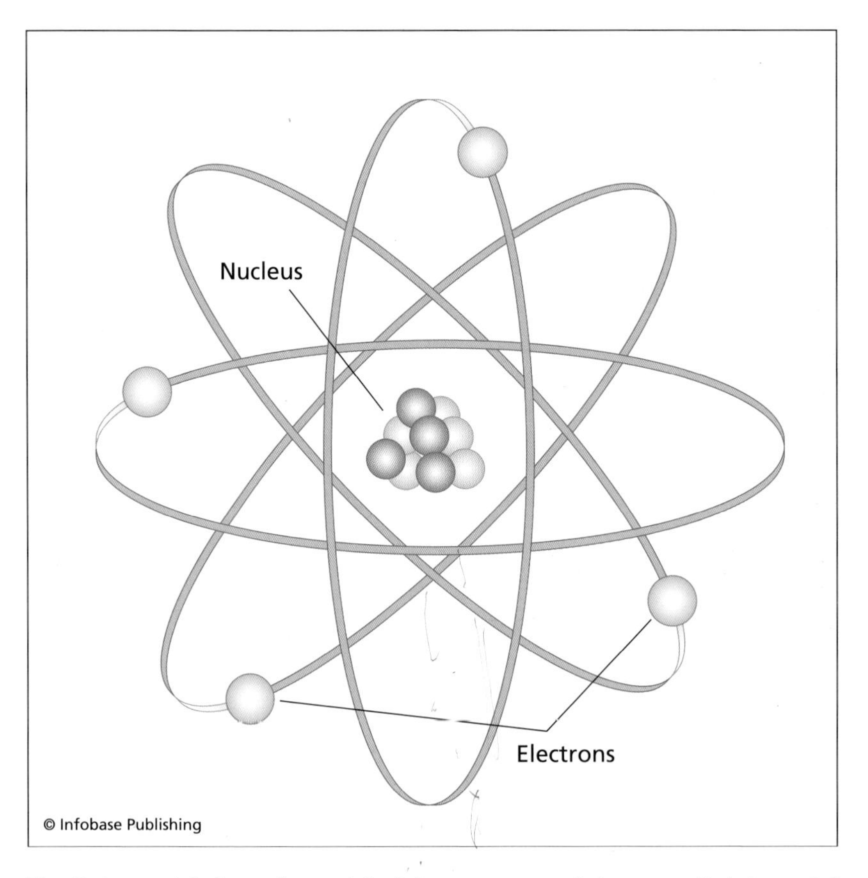

The Bohr model. An early model of the geometry of the atom, Bohr's model envisioned atoms as solar systems in miniature.

and with the exception of some hydrogen atoms all atoms contain at least one neutron. All of an atom's protons and neutrons are clustered tightly together in the nucleus. Neutrons and protons are of roughly equal masses, and by atomic standards they are, indeed, very massive: A single proton or neutron is approximately 1,800 times as massive as an electron.

In addition to their masses, these three constituents of an atom are also characterized by their electrical charges. Each proton carries one positive electrical charge; each electron carries one negative electrical charge. Although these charges are opposite in sign, they are exactly equal in magnitude. Scientists have assigned a single positive charge the value of +1 and a single negative charge the value of -1. Because each atom contains precisely as many protons as electrons—that is, each atom contains exactly as many positive electrical charges as negative ones—atoms are said to be electrically neutral; the sum of all the positive and negative charges equals zero. Neutrons, which comprise the third component of an atom, are electrically neutral; that is, they carry no net electrical charge. It is possible, however, for a neutron to change into a proton. This is part of a process called transmutation: A neutron located within a nucleus emits an electron, called a beta ray or *beta particle,* and what remains behind is a proton. In a nuclear reactor the emission of beta particles is a vital part of several very important processes.

Atoms are classified by the number of protons present in the nucleus. This number is called the *atomic number* of the atom. By way of example, all hydrogen atoms have exactly one proton, and so they are said to have an atomic number of 1; all helium atoms have exactly two protons, and they have an atomic number of 2. The atomic number of an atom can always be obtained from the periodic table of elements. Each box in the periodic table represents one element. Each element is identified by its own unique abbreviation—the letter H, for example, stands for hydrogen, and the letter O stands for oxygen—and the number in the upper left corner of each box is the atomic number corresponding to that atom.

But not all atoms of the same chemical element are identical, even though all such atoms have the same atomic number and consequently the same number of protons and electrons. If two atoms with the same atomic number are different, the difference occurs in their nuclei. They possess different numbers of neutrons. Atoms with the same atomic number but different numbers of neutrons are said to be different *isotopes* of the same element. For example, more than 99 percent of all naturally occurring uranium atoms (atomic number 92) are composed of atoms containing 146 neutrons within their nuclei. Almost all of the remaining naturally occuring uranium atoms contain 143 neutrons.

Scientists have developed a shorthand system of notation to identify different atoms and different isotopes of the same atom. The system consists of the abbreviation of the name of the element together with a subscript and a superscript. The subscript gives the atomic number of the atom—that is, the subscript identifies the number of protons present in the nucleus—and the superscript identifies the total number of protons *and* neutrons present in the nucleus. This is called the *mass number.* By way of example, ${}^{238}_{92}U$ represents the isotope of uranium that has 146 neutrons in its nucleus. (*U* is the scientific abbreviation for uranium, 92 is the number of protons in the nucleus of every uranium atom, and 238 is the mass number. Notice that 238 minus 92 equals 146, the number of neutrons in this isotope.) With this notation one can describe any isotope of any atom. But in order to describe how the nuclei of atoms split and why so much energy is released in the process, other concepts are needed.

CONSERVATION LAWS AND RADIOACTIVE PROCESSES

Physical processes are often described in terms of *conservation laws.* A conservation law is a statement that a particular property—energy, for example—can neither be created nor destroyed. A hypothetical process that violates a conservation law *cannot* occur, not in the laboratory and not in nature. Conservation laws are valuable because they assist scientists in separating what is true from what is false. Because these laws of nature are so important, a great deal of scientific research is devoted to the discovery of new conservation laws and to the application of previously established ones. In the field of nuclear physics, which is that branch of study concerned with atomic nuclei, scientists have discovered a number of conservation laws. The following three conservation laws are especially important in understanding reactions that involve the nuclei of atoms:

1. The total number of protons and neutrons participating in a nuclear reaction is conserved—that is, the sum of all protons and neutrons is the same before, during, and after a nuclear reaction. This is often summarized by saying that the mass number is conserved.
2. The total electrical charge—that is, the sum of all positive and negative charges in the atoms and any other particles participating in a nuclear reaction—is conserved. (Recall that a single positive charge is assigned a value of +1 and a negative charge is assigned a value of -1, and the total charge is the sum of the individual charges.)
3. The total energy of the atoms and other particles involved in a nuclear reaction is conserved.

For purposes of this book, the term *other particles* in the third conservation law means the protons, neutrons, and electrons that under certain circumstances are not bound to an atomic nucleus. The word *conserved* that appears in each of the preceding statements means that the quantity under consideration is the same before, during, and after a reaction has taken place. During the process of transmutation described earlier, for example, when a neutron, which has zero electrical charge, produces an electron, or a beta particle, which has a charge of -1, what remains behind is a proton, which has an electrical charge +1. Add the charge of the beta particle to the charge of the proton and the result is zero electrical charge, which was the charge value of the neutron.

To illustrate the use of the notation developed in the preceding section and the way that a conservation law can be applied, consider the isotope of uranium containing a total of 235 protons and neutrons. It is often called uranium-235, and it is often written as

$^{235}_{92}U$. The nucleus of uranium-235 is not entirely stable; sometimes the nucleus of this isotope will spontaneously split apart. This splitting can happen in a variety of ways. Sometimes the nucleus of $^{235}_{92}U$ spontaneously breaks, or splits, into two pieces, and in the process produces an isotope of the element thorium whose symbol is $^{231}_{90}TH$. (Thorium is a metallic element often found in conjunction with uranium ore.) As its symbol indicates, thorium has an atomic number of 90, and this particular isotope of thorium contains 231 protons and neutrons, which is just another way of saying that this isotope has 141 neutrons (231 - 90 = 141). In addition to the thorium atom, the spontaneous splitting of $^{235}_{92}U$ also produces something called an *alpha particle.* The alpha particle has a mass number of four.

Given the fact that exactly two particles were produced when the $^{235}_{92}U$ atom split, if it is known that one particle was a thorium-231 atom, the mass number of the second particle, the alpha particle, could have been predicted. Here is how: Because the uranium atom had a mass number of 235, the sum of the mass numbers for the thorium-231 atom and the alpha particle that resulted from this splitting of $^{235}_{92}U$ must also equal 235, a consequence of the statement that mass number is conserved. (See conservation law number 1 on page 28.) The thorium atom has a mass number of 231, so the mass number of the alpha particle must be 4 (235 = 231 + 4). This particular process of splitting $^{235}_{92}U$ into $^{231}_{90}Th$ and an alpha particle can be described in the following way:

$$^{235}_{92}U \rightarrow {}^{231}_{90}Th + {}^{4}_{2}He$$

The symbol $^{4}_{2}He$ represents the alpha particle, which is nothing more than the nucleus of a helium atom. (The symbol for helium is He.)

It is the mass number—not the atomic number—that is fundamental for these kinds of processes. In many nuclear reactions the atomic number fails to be conserved, but the mass number is always conserved. It is a law of nature.

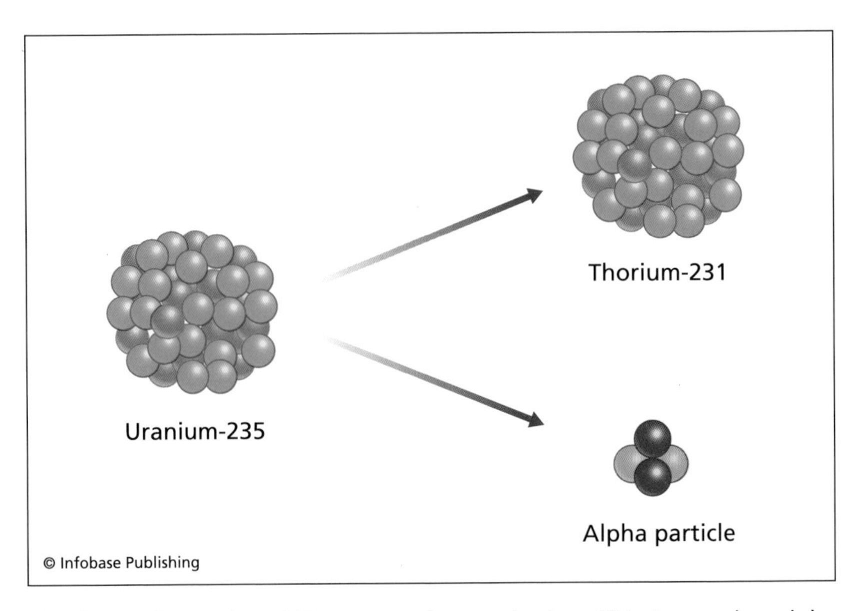

The decay of a uranium-235 atom produces a thorium-231 atom and an alpha particle

Isotopes of many different types of atoms undergo spontaneous changes in ways that are similar to the spontaneous splitting of ${}^{235}_{92}U$ described in the preceding paragraph. When such an event occurs, particles—electrons or neutrons, for example—and gamma rays, which are high-energy electromagnetic waves, may radiate outward from the splitting nucleus. These events are all part of the process of radioactive decay, a process that occurs naturally and continually throughout the environment.

Isotopes that undergo radioactive decay are said to be radioactive. In one sense, radioactive decay is a random phenomenon, which is to say that there is no way of predicting when a particular radioactive atom will undergo the process of decay. In might happen during the next second, or it might not happen for a billion years. Much depends on the particular isotope under consideration, but no matter how much is known about any particular isotope, the timing of the actual decay event cannot be predicted. Scientists have discovered, however, that given a large sample of atoms, all of which have identical atomic numbers and identical mass numbers, it is possible to predict with a great deal of precision the length of time that must elapse until only half the sample remains. (During this time the other half of the original sample will have "decayed" into other types of atoms.) This interval of time is called the half-life of the isotope.

Even among the isotopes of a single element, half-lives vary tremendously. For example, the half-life of krypton-85, written $^{85}_{36}Kr$, is about 11 years, but krypton-87, written $^{87}_{36}Kr$, has a half-life of only about 76 minutes. Mathematically, the problem of predicting the time to decay of an individual atom is similar to the problem of predicting the outcome of a coin flip: While no one can predict the outcome of a single flip of a coin with more than 50 percent accuracy, it is, nevertheless, a near certainty that after flipping a single coin many times, almost precisely 50 percent of the flips will result in an outcome of heads. Furthermore, the more often one flips the coin, the more confident one can be in the frequency of heads that one will observe. Accurately predicting the half-life of a particular sample of radioactive material or the frequency of heads in a long series of flips is relatively easy, but accurately predicting the timing of individual events is as impossible for radioactive atoms as it is for coin flips.

But if radioactive isotopes are continually falling apart, why are there any radioactive isotopes left in the environment? Earth is, after all, 4.5 billion years old. It might seem that any nucleus of any element that could have decayed would have already decayed during such a lengthy period of time. There are two answers to this question. First, some radioactive isotopes are produced naturally and continually. Carbon-14, also written as $^{14}_{6}C$, a radioactive form of the element carbon, is produced naturally in Earth's atmosphere due to the interaction of cosmic rays with nitrogen-14 ($^{14}_{7}N$) atoms. Carbon-14 has a half-life of "only" 5,730 years. Any carbon-14 present on Earth when the planet was initially formed has long since decayed, but because small amounts of $^{14}_{6}C$ are continually produced, some of this isotope is still present in the environment.

By contrast with carbon-14, other isotopes of other atoms are not produced naturally. The isotope $^{238}_{92}U$ is not, for example, the result of any natural process within Earth's environment, and because it is radioactive, atoms of $^{238}_{92}U$ are continually self-destructing. Consequently, the amount of $^{238}_{92}U$ present on Earth today is simply the total amount of $^{238}_{92}U$ initially present on Earth when the planet was first formed minus the amount that has decayed over the entire course of Earth's history. In the case of $^{238}_{92}U$, which has a half-life of about 4.5 billion years, the amount currently present is about half of what was present originally because uranium-238's half-life is roughly equal to the age of Earth. There are other isotopes of other elements that simply do not occur on Earth any longer because they are not produced by any natural process, and their half-lives are so short compared to the age of the planet that the material originally present at the planet's formation has long since self-destructed. In order to observe these relatively short-lived materials, they must be produced in laboratories.

Finally, as radioactive isotopes decay, they emit energy. To appreciate how much energy is emitted, it is important to bear in mind this important, curious, and remarkable fact: The principle of conservation of mass does *not* apply to nuclear reactions. The principle of conservation of mass states, in part, that the mass or quantity of materials present before a chemical reaction takes place equals the mass of the materials present during and after the chemical reaction is completed. Chemists and chemical engineers make frequent use of this conservation law when they seek to understand a new chemical reaction. But nuclear reactions do not conserve mass. (The difference between a chemical and a nuclear reaction is that nuclear reactions involve making and breaking the very powerful bonds that bind together the protons and neutrons within an atomic nucleus. By contrast, chemical reactions involve making and breaking the much weaker bonds that bind atoms together within a molecule.)

Scientists have performed a number of sophisticated experiments to measure the mass of an isolated proton and an isolated neutron. These measurements are extremely accurate. But when a group of protons and neutrons form an atomic nucleus, the mass of the nucleus will be *less* than the sum of the masses of its constituent parts. What happened to the missing mass? In the process of creating the nucleus, some of the mass of the particles in the nucleus was converted into the energy that binds the protons and neutrons of the nucleus together. This "missing mass phenomenon" illustrates the fact that mass and energy can be converted one into the other. The relationship between mass and energy is remarkably simple to state and is given by Albert Einstein's most famous equation:

$$E = mc^2$$

where the letter E represents energy, m represents mass, and c represents the speed of light. (Pronounced "c squared," c^2 is shorthand for $c \times c$).

Einstein's equation also explains why nuclear reactions have the potential to release so much more energy than chemical reactions—why, for example, a thimbleful of nuclear reactor fuel can release as much energy as can be obtained by burning a million times as much coal. (Burning is a chemical reaction.) The speed of light is very fast, so by squaring it—that is, by multiplying c by c—one obtains a very large quantity indeed. Therefore, even a small amount of mass, when multiplied by the very large quantity c^2, represents a very large amount of energy. When an atomic nucleus is split, some of this binding energy is released and appears in the form of thermal energy. The goal of a nuclear reactor is to release this thermal energy in a controlled way and then to harness it to produce the electricity on which all modern societies depend. To understand how this occurs and the consequences of producing thermal energy in this way, one must understand a little about the nuclear reactions by which this energy is obtained.

ATOMIC FISSION

Nuclear reactors are designed as places where the nuclei of atoms can be split in a controlled way in a process called nuclear fission. The heat that is released during fission is then harnessed to do work. The splitting of a single atom does not release enough energy to generate usable amounts of electricity, but to understand the process of fission, the splitting of one atom is a good place to start.

When a neutron collides with the right type of atom, the nucleus of the atom may absorb the neutron. The additional energy that the neutron transfers to the nucleus changes the balance of forces holding the nucleus together. Sometimes this extra energy is enough to break apart the nucleus. Gamma rays may radiate out of the excited

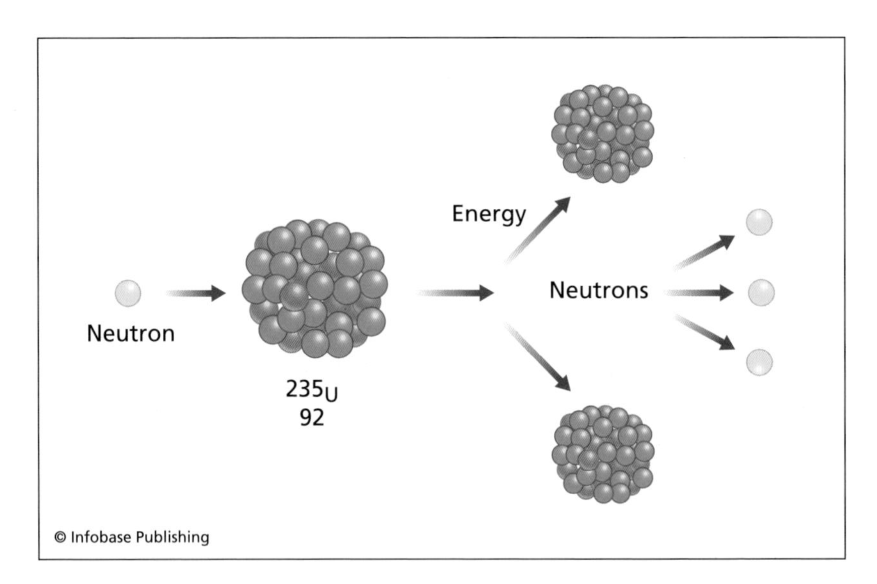

Atomic fission. One free neutron gives rise to fission products including (sometimes) three free neutrons, thereby making a chain reaction possible.

nucleus, and one or more neutrons may be ejected from the nucleus as it splits into two or more smaller atoms. (Here the word *smaller* means that each of these newly formed atoms has a mass number that is smaller than that of the original parent atom.) This process is called fission. The fission process transfers energy to the surrounding material, and some of that additional energy appears in the form of thermal energy.

To illustrate the process of fission, consider the following typical fission reaction. In this reaction, a neutron collides with a uranium-235 atom and creates a highly unstable uranium-236 atom. Almost as soon as the uranium-236 atom is created, it splinters into an isotope of barium ($^{144}_{56}$BA), an isotope of krypton ($^{89}_{36}$Kr), and three free neutrons are produced as well. This process can be summarized using the notation that was introduced earlier (the *n* in the equation stands for the word "neutron"):

$$n + {}^{235}_{92}U \rightarrow {}^{236}_{92}U^{*} \rightarrow {}^{144}_{56}BA + {}^{89}_{36}Kr + 3n \qquad \textbf{(2.1)}$$

The asterisk attached to the $^{236}_{92}$U is there to indicate that the atom is in an excited state and exists only for a tiny fraction of a second. Notice, again, that mass number is conserved: On the left side of equation (2.1) there is the uranium-235 atom plus the one additional neutron that initiates the reaction for a total mass number of 236. Together they produce the highly unstable $^{236}_{92}$U* atom. On the right side of equation (2.1), the barium atom has a mass number of 144, the krypton atom has a mass number of 89, and 3 free neutrons are produced. Add these three numbers together and one again obtains the sum of 236, the same mass number as on the left side of the equation. Notice also that in this reaction one free neutron went in and three free neutrons came out. The three neutrons that are produced in the reaction can potentially cause three additional fissions. This explains how one can begin with a few random fission events and quickly produce an enormous increase in thermal energy output. Controlling the energy output resulting from this type of fission process is a major challenge for nuclear engineers.

In a nuclear reactor neutrons are careening around the fuel at widely varying speeds and striking different nuclei at random. It should not be surprising, therefore, to learn that many different atoms are produced as a result of the fission process, and this is exactly what happens. Furthermore, many fission products—that is, many of the nuclei that are the result of nuclear fissions—are themselves radioactive and so begin the process of radioactive decay as soon as they are formed. In the process of decay they give rise to still other types of atoms. In fact, many elements are created inside reactors, either during the fission process or as a consequence of the radioactive decay of the products of fission. This is important because it reveals something crucial about how thermal energy is produced within the reactor: While splitting the nuclei releases a great deal of energy per event, it is only the start. The products of fission continue to release energy as they decay.

Heat generation only stops when the process of fission ceases *and* when the radioactive isotopes finally decay to stable nonradioactive isotopes. Depending on the atoms that are initially produced by a particular fission event this can take minutes, days, weeks, centuries, or millennia, and at each step of the radioactive decay process energy is released. This explains why one cannot turn off a nuclear reactor the way that one turns off a lightbulb. While there are technologies that enable one to quickly "soak up" the free neutrons within the reactor and so end the process of nuclear fission, the products of the fission reaction continue to generate thermal energy as a result of radioactive decay. There is no way to stop this process.

As a practical matter, however, once the fission process has stopped, the production of thermal energy drops off fairly quickly. If a reactor has been operating at full power for at least a month—so that many fission reactions have already occurred within the fuel—the reactor will initially generate as much thermal energy

Fermi's Reactor

"Some recent work by E. Fermi and L. Szilard, which has been communicated to me in manuscript, leads me to expect that the element uranium may be turned into a new and important source of energy in the immediate future."

—*Albert Einstein, in a letter sent to President Franklin D. Roosevelt*

During World War II the United States undertook a secret program, code named the Manhattan Project, to build a nuclear weapon. Some of the nation's best engineers and scientists were assigned to this project. One of them was the Italian-born American physicist Enrico Fermi (1901–54).

Prior to the start of World War II, a number of scientists, including Fermi, Niels Bohr, and Hungarian-born physicist Leo Szilard (1898–1964) had come to believe that a nuclear weapon was possible. They based their belief, in part, on a series of experiments that had been performed earlier in Germany. German physicists had discovered that when uranium atoms are bombarded with neutrons, the elements barium, krypton, and some additional neutrons are produced. (This is the nuclear reaction described in equation [2.1].) The fact that one neutron initiated the reaction and multiple neutrons were produced suggested the possibility of a self-sustaining reaction, a so-called chain reaction, because each of the product neutrons could cause an additional reaction. Roughly speaking, the pattern would be one neutron produces three, then three produce nine, then nine produce 27, then 27 produce 81, and so on. Because individual reactions happen so fast, this type of reaction could produce a tremendous amount of energy almost instantly.

Worried that Germany was also hard at work on developing such a bomb, the United States devoted a great deal of its resources to the Manhattan Project. Fermi was assigned the task of building a device that would produce a controlled chain reaction. This was the world's first

(continues)

(continued)

nuclear reactor—he called it an atomic pile—and it was built on a squash court at the University of Chicago.

Fermi's reactor really was a pile, a pile of graphite bricks. He knew that when neutrons are first released during a fission event, they are traveling so fast that they are unlikely to be absorbed by uranium-235 atoms, and unless they are absorbed no fission will result. In order to increase the chance that the neutrons produced by a random fission will be absorbed by still other uranium-235 atoms, the neutrons needed to be moderated or slowed down. Graphite, Fermi knew from previous experiments, is a good neutron *moderator,* and so when it came time to build the pile, Fermi used 380 tons of graphite blocks and almost 50 tons of uranium. The uranium was loaded into some of the graphite blocks. Other blocks contained no uranium. The blocks were then piled in alternating layers: One layer of blocks contained inserts of uranium; the next layer was composed of blocks of pure graphite. The pile was 57 layers high and braced with wooden beams. *Control rods,* long rods covered with cadmium, a material that absorbs neutrons the way a sponge absorbs water, were inserted into the pile to prevent a spontaneous chain reaction. On December 2, 1942, when all preparations were complete, the control rods were slowly withdrawn, and the first controlled chain reaction began. It was allowed to continue for 28 minutes as Fermi and his coworkers made measurements to verify their results. When they were satisfied, the control rods were reinserted, and one of the most important experiments of the 20th century was brought to a conclusion.

immediately after shutdown as it did when it was operating at full capacity, but by the end of the day its thermal energy output will have dropped by about 90 percent. Although this is a large drop, nuclear reactors generate so much thermal energy that 10 percent of the total thermal output still represents a great deal of energy.

A classified patent was filed with the U.S. patent office in December 1944 for this first atomic pile. It listed Fermi and Szilard as coinventors. The patent was issued on May 18, 1955, six months after Fermi's death.

Fermi's atomic pile was used to create the first controlled self-sustaining fission reaction. (Shown with Dr. Henry W. Newson.) *(Argonne National Laboratory, AIP Emilio Segrè Visual Archives)*

When a neutron collides with a nucleus, other outcomes besides fission are possible. Two other outcomes are especially important in understanding the way that nuclear power is generated. First, the neutron may simply careen off the nucleus of the atom with which it collides. Some of the energy of motion of the neutron will be transferred to the nucleus that it strikes, and as a consequence the atom will move somewhat faster than it did prior to the collision. The neutron will move away from the collision site more slowly than it approached. The reason for the slowing of the neutron can be found in conservation law number three, listed earlier in the chapter: Energy—in this case the energy of motion, also called *kinetic energy*—is conserved. In other words, the kinetic energy that was initially carried entirely by the neutron is, after the collision, shared between the neutron and the nucleus with which it collided. Consequently, the neutron will have less kinetic energy after the collision than it did before the collision took place. The collision will have moderated, or slowed, the speed of the neutron.

There is one more outcome that can result when a neutron collides with a nucleus. The nucleus will absorb the neutron without breaking apart. As with a fission event, the absorption of a neutron generally starts a chain of events that can take hours, days, weeks, or sometimes many years before it ends in the production of a stable nucleus.

To illustrate this phenomenon consider the following very important reaction. It occurs in all commercial power reactors. A free neutron collides with an atom of uranium-238. It is absorbed by the $^{238}_{92}U$ atom to produce a new isotope of uranium, $^{239}_{92}U$. (When the neutron is absorbed, the new isotope emits energy in the form of a gamma ray.) In symbols:

$$n + {}^{238}_{92}U \rightarrow {}^{239}_{92}U + (gamma\ ray) \qquad \textbf{(2.2)}$$

The new isotope, uranium-239, is not very stable. It has a half-life of about 23 minutes. It soon emits energy from its nucleus in the form of an electron, called a beta particle, and one of the neutrons within the nucleus is changed into a proton. As was described earlier in the chapter, this process, called transmutation, is governed by the second conservation law, which states that electrical charge is conserved—that is, the creation of an electron must be accompanied by the creation of a proton. In this way, the total electrical charge of the system of particles remains the same at the end of the process as it was at the beginning. The result is a new element with atomic number 93. The new element is called neptunium, and its symbol is $^{239}_{93}Np$. In symbols, this transmutation reaction is written as follows:

$$^{239}_{92}U \rightarrow {}^{239}_{93}Np + (beta\ ray) \quad \textbf{(2.3)}$$

Neptunium-239 is also radioactive. It has a half-life of about 56 hours, and when it decays, an electron is ejected from its nucleus as another neutron is changed into a proton. The result of this transmutation reaction is an isotope of plutonium, $^{239}_{94}Pu$:

$$^{239}_{93}Np \rightarrow {}^{239}_{94}Pu + (beta\ ray) \quad \textbf{(2.4)}$$

Plutonium-239, a toxic silver-colored element, is also radioactive, but its half-life is about 24,000 years. As a consequence of the very short half-lives of neptunium-239 and uranium-239, when the fuel is finally removed from the reactor, one finds very little $^{239}_{92}U$ and very little $^{239}_{93}Np$ but a good deal of $^{239}_{94}Pu$. The element plutonium, which does not occur naturally on Earth, was created entirely within the reactor. Moreover, plutonium-239 is *fissile*—that is, it can undergo fission—and so can be used as a reactor fuel. Consequently, a reactor that uses a fuel containing $^{238}_{92}U$ is not just consuming fuel, it is *creating* fuel in the form of $^{239}_{94}Pu$. There are even reactors that create more fuel, in the form of plutonium, than they consume in the form of uranium-235. They are called breeder reactors.

As a practical matter, the process of fission described in this chapter will only occur for very specific types of materials. Consider the example of the fission of $^{235}_{92}U$ given earlier in this section: When the single neutron collided with uranium-235, three neutrons together with some other reaction products were produced. Among all the isotopes of all the atoms in the periodic table, uranium-235 is unusual in the sense that it can be split by a neutron of moderate energy and produce more than one free neutron when it undergoes fission. By careful study of uranium-235, scientists have found that in the conditions that prevail within a commercial nuclear reactor, every 10 fission events produce an average of 25 free neutrons. This is usually expressed by saying that uranium-235 produces an average of 2.5 neutrons per fission, although no fission can produce just half of a neutron.

Although it might seem that it is only necessary to produce one free neutron per fission to create a chain reaction, in practice surplus neutrons are critical for sustaining the chain reaction because some neutrons always escape the fuel without causing additional reactions, and some neutrons will be absorbed by atoms within the fuel without causing a fission reaction. Materials such as uranium-235 that produce enough neutrons per fission event to sustain a nuclear chain reaction are called fissile.

Again, most elements are poorly suited for establishing and maintaining nuclear chain reactions, because their nuclei are too difficult to split. Even uranium-238, which only differs from uranium-235 by the presence of three additional neutrons, will not sustain a fission reaction. As previously described, it absorbs the colliding neutron without splitting, but in the process it initiates a sequence of events that results in the production of fissile material, plutonium-239. Materials such as uranium-238 that can contribute to a chain reaction without being able to sustain the reactions themselves are called *fertile.* Just as most isotopes of most elements are not fissile, neither are they fertile. Most atoms have nuclei that are too stable to serve as reactor fuel.

Nuclear reactor fuel is carefully crafted to contain both fissile and fertile materials. A fuel consisting solely of fissile materials can release too much energy too quickly, and a nuclear explosion is the result. The fuel in a nuclear reactor cannot cause a nuclear explosion. Nuclear fuel is constructed in a way that facilitates the production of thermal energy. It does not have the composition required to make a bomb.

Reactor Fuel

Uranium is the main component of fuel for commercial nuclear reactors, but the way that the uranium is processed for insertion into fuel assemblies depends very much on the type of reactor in which the fuel is to be used. These details are important because nuclear reactor fuel is unlike other fuels. It releases far more energy per unit mass than any fossil fuel, and it contains the same materials needed to make nuclear weapons. In order to understand nuclear power, it is important to learn something about nuclear fuel, and that is the goal of this chapter.

MANUFACTURING REACTOR FUEL

The process of manufacturing nuclear fuel often, but not always, begins in uranium mines. (Fuel can also be recycled, but for reasons described later in the chapter, some countries have made a conscious decision not to recycle their fuel.) Uranium is not difficult

Cigar Lake Uranium Mine near Saskatoon, Canada, location of one of the world's richest deposits of uranium ore *(Cameco)*

to find. It exists throughout the environment at very low concentrations. Whether a particular uranium deposit contains the metal at concentrations sufficient to make mining worthwhile depends on the price of uranium. At current prices, most economically viable uranium deposits are located in Algeria, Australia, Brazil, Canada, Namibia, Niger, South Africa, and the United States. Uranium is also present in low concentrations in seawater, and Japanese scientists have extracted significant amounts of uranium from seawater but not at a cost that is competitive with conventionally mined uranium. If the cost of uranium increases, or the cost of seawater extraction decreases, the oceans might also be profitably mined for uranium. While uranium concentrations in seawater are low, the total amount of uranium present is very large because the oceans' capacity is huge.

Once the uranium has been mined and the uranium ore separated from the other material with which it was found, it is ready for the manufacturing process. The next step in the process depends upon the design of the reactor in which the fuel is to be used. There are different types of reactors, each with its own fuel requirements. Only three types of commercial reactors will be considered in this chapter: boiling water reactors and pressurized water reactors, which are collectively called light water reactors (LWRs), and the class of reactors called CANDU reactors, short for Canadian deuterium-uranium reactors. Together, these three types generate more than 90 percent of all the nuclear-generated electricity throughout the world. The designs of all three types of reactors will be described in more detail in chapter 4; in this chapter only the characteristics of the fuel on which they depend are described.

All LWRs use what is called slightly *enriched uranium*. Natural uranium is composed of 99.3 percent uranium-238 and 0.7 percent uranium-235. Slightly enriched uranium contains a higher proportion of uranium-235 than natural uranium. The proportion of uranium-235 is increased to between 2 and 5 percent of the total; the remainder of the slightly enriched uranium consists of the more common isotope, uranium-238. In contrast to LWRs, CANDU reactors generally use natural uranium for their fuel. They are, however, more flexible than LWRs with respect to their fuel in that they can also run on slightly enriched uranium.

To appreciate how reactors use their fuel, it is helpful to take a closer look at the way that fission actually occurs inside a reactor. When an atom of ${}^{235}_{92}U$ undergoes fission and free neutrons are ejected into the surrounding fuel, the just-released neutrons are traveling at very high velocities. These high-energy neutrons are, not surprisingly, called *fast neutrons*, and they travel at speeds in the neighborhood of 6,200 miles per second (10,000 km/s)—but only over very short distances, at which point they collide with nearby atoms. Fast neutrons are not readily absorbed by $^{235}_{92}U$, and as a consequence fast neutrons produce very few fissions. By contrast, slow neutrons, also called *thermal neutrons,* which have speeds in the vicinity of 1.4 miles per second (2.2 km/s), are much more likely to be absorbed by $^{235}_{92}U$ atoms. If the fast neutrons are moderated, or slowed, to thermal speeds, the number of fissions increases dramatically. The successful operation of a reactor, therefore, depends on the designer's ability to control the speed of the free neutrons. (Thermal speeds constitute a lower speed limit for neutrons. All atoms in the fuel are in constant vibratory motion. The speeds at which they vibrate are related to the temperature of the material of which they are a part: the faster the vibrations, the higher the temperature. At thermal speeds the neutrons are in equilibrium with the atoms that surround them in the sense that they jostle the surrounding atoms as much as the surrounding atoms jostle them.) The method by which reactor designers moderate fast neutrons is one of the most important design decisions that nuclear engineers must make, because it determines the characteristics of the fuel that can be used in the reactor.

LWRs use ordinary water to moderate the fast neutrons. Recall that a water molecule consists of one oxygen atom bound to two hydrogen atoms. During the process of neutron moderation a fast neutron repeatedly collides with the hydrogen atoms in the surrounding water molecules. With each collision the neutron transfers some of its energy of motion, also called kinetic energy, to the hydrogen atom with which it collided. In the process the hydrogen atom speeds up somewhat and the neutron slows down a little. In this way neutrons make the transition from being fast neutrons to being thermal neutrons. The phenomenon is similar to what happens when one rapidly rolls a single billiard ball onto a billiard table on which many balls have been randomly positioned. With each collision the rapidly rolling ball transfers some of its energy of motion to one of the other balls. These balls begin to roll about the table and collide with one another. After several collisions, the kinetic energy, or energy of motion, of the original ball is distributed among the other balls that were positioned on the table, and all the balls roll with the same average speed.

The analogy between billiard balls and neutrons is not perfect, however, because billiard balls soon stop rolling because of friction between the balls and the table. This does not happen at the atomic level. The random vibrations experienced by the molecules inside the reactor are of a more long-lasting nature. If, after having been moderated, a neutron happens to reenter the *fuel rod* and encounters an atom of $^{235}_{92}U$, it is much more likely to initiate a fission reaction.

Ordinary water is used to moderate fast neutrons in LWRs because it is plentiful and cheap, and its properties at different temperatures and pressures are fairly well understood, but it is not an ideal moderator. The problem associated with using ordinary, or light, water is that the isotope of hydrogen that is found in almost all molecules of light water—the isotope that contains no neutrons in the nucleus—sometimes absorbs neutrons. When a neutron is absorbed, it is no longer unavailable to initiate new fission reactions. The percentage of free neutrons absorbed via collisions with hydrogen atoms is not large but taken together with other losses—some neutrons, for example, are absorbed by the isotope $^{238}_{92}U$—enough neutrons escape or are otherwise absorbed to prevent a chain reaction from developing. In fact, if natural uranium were used as a reactor fuel inside a reactor that used ordinary water as a moderator, a chain reaction would not occur.

Designers of LWRs solve the problem of neutron loss by changing the isotopic composition of the uranium fuel. Recall that natural uranium consists almost entirely of the isotope $^{238}_{92}U$. Almost all of the remainder—0.7 percent—consists of $^{235}_{92}U$. Or to put it another way: In a sample of natural uranium, approximately one out of every 140 uranium atoms is an atom of uranium-235. One can increase the supply of free neutrons by increasing the percentage of $^{235}_{92}U$ atoms present in the fuel. Changing the isotopic composition of the uranium fuel in this way increases the number of spontaneous fissions per unit time. Finding a way to enrich uranium was one of the great technical achievements of the Manhattan Project, the secret World War II project undertaken by the United States that led to the invention of the atomic bomb.

Several methods now exist to enrich uranium. For purposes of illustration, a common three-step method is described here. It involves using high-speed gas *centrifuges,* devices that spin at very high speeds for extended periods of time. The procedure works as follows: First the uranium is combined with the element fluorine to produce the chemical compound uranium hexafluoride, each molecule of which consists of a single uranium atom bound to six fluoride atoms. Its chemical formula is UF_6. Engineers use UF_6 because it exists as a gas at temperatures and pressures that can easily be maintained during the enrichment process.

Next, the centrifuge is filled with the gas and spun very rapidly. The ever-so-slightly heavier molecules of UF_6, the ones formed from $^{238}_{92}U$, tend to concentrate in the outer areas of the cylinder, leaving the slightly lighter molecules, the ones formed from $^{235}_{92}U$, nearer the center of the centrifuge. As one might expect, this process is not very efficient. The difference between the two types of uranium hexafluoride molecules is just the mass of three neutrons, so the effect is very weak. But the effect exists, and engineers and scientists have learned to exploit it. Gas extracted from the region where one would expect to find more $^{235}_{92}U$ atoms is slightly richer in this isotope.

The gas in the region richer in $^{235}_{92}U$ is extracted from the cylinder, and the procedure is repeated again. Once the procedure has been repeated a few thousand times, the percentage of $^{235}_{92}U$ present in the UF_6 will have reached the desired level of enrichment—the rest of the uranium is still $^{238}_{92}U$—and the uranium is returned to a solid form through a series of chemical reactions. Reactor fuel constructed from this enriched uranium produces more free neutrons than are produced from natural, unenriched uranium fuel. Consequently, when fast neutrons are moderated inside LWRs using light water, there are sufficient "extra" neutrons available, despite the absorption of some neutrons by the hydrogen atoms, to begin and maintain a nuclear chain reaction.

In contrast with LWRs, CANDU reactors use what is called "heavy water" to moderate the fast neutrons generated by the fission reaction. Molecules of heavy water consist of one oxygen atom paired with two deuterium atoms. (Recall that deuterium is the isotope of hydrogen with a mass number of two.) The advantage of using heavy water is that unlike light water, deuterium atoms rarely absorb an impinging neutron. Consequently, in a CANDU reactor, very few neutrons are lost to the moderator, so there is no need to enrich the reactor fuel. CANDU reactors work efficiently using uranium fuel in its natural isotopic composition. The disadvantage of heavy water lies in its high cost. Heavy water must be manufactured, and to be an effective moderator it must be very pure. CANDU reactors use a lot of heavy water, but the ability to use natural uranium as fuel, thereby avoiding the enrichment process, is a powerful incentive, and a number of countries use these reactors.

Perhaps the single most important consequence of the decision to use enriched uranium as fuel lies not in the fuel itself but in the technology needed to produce it. The same technology that is used to produce slightly enriched uranium can be used to produce highly enriched uranium, and highly enriched uranium can be used to build nuclear weapons. Therefore, those countries that possess enrichment technology can use it to produce reactor fuel or nuclear weapons. Controlling access to enrichment technology has important consequences in the drive to control the spread of nuclear weapons.

INSIDE THE REACTOR

Uranium fuel cannot be simply dumped inside the pressure vessel. The way that the fuel is arranged—the fuel geometry—is important for five reasons. First, fuel geometry should facilitate the moderation of fast neutrons. Second, it is easier to control the fission rate with the right fuel configuration. Third, the fuel geometry must enable the coolant to circulate easily in order to transfer the thermal energy from the fuel, where it is produced, to the working fluid, where it is harnessed to do work. Fourth, the reactor fuel must be packaged in such a way that the products of fission, which are often highly radioactive, remain safely contained. Finally, reactors require many tons of fuel to produce the requisite amount of power. This fuel must be packaged so as to facilitate its insertion and withdrawal from the reactor. LWRs, for example, are shut down periodically for refueling. Replacing the fuel quickly helps to minimize the time that the reactor is not producing power, an important economic consideration.

After fuel enrichment, the next step in the process of manufacturing fuel for LWRs—CANDU reactor fuel is described in more detail later in this section—involves converting the enriched uranium, which is still in the form of the gas UF_6, back into solid form. The fluorine atoms are separated from the uranium atoms, and the uranium atoms are combined with oxygen atoms to produce the compound uranium dioxide, whose chemical symbol is UO_2. The compound UO_2 is produced for use inside the reactor. Designers favor this compound because uranium dioxide is a solid at room temperatures, and it remains a solid at the high temperatures at which a nuclear reactor normally operates. It is also resistant to radiation damage.

The uranium dioxide is shaped into small cylindrical pellets. The exact size of the pellets depends on the details of the particular LWR, but they are roughly similar across different designs. By way of example, in one Westinghouse-designed pressurized water reactor, the pellets are roughly one-third of an inch (.8 cm) in diameter and about one-half inch (1.35 cm) in length. They are loaded into long, narrow tubes composed of a material called zircaloy, which is chosen because it absorbs relatively few neutrons and because it is strong and resistant to radiation damage. For a LWR, the tubes, called fuel rods, are generally fairly long. In the Westinghouse design used as an example here, the fuel rods are about 12 feet (3.7 m) long. The inside of the tube is slightly larger than the pellet in order to accommodate the expansion of the pellet as its temperature increases. The space between the pellet and the inside of the fuel rod is filled with helium to facilitate the movement of thermal energy from the pellet to the fuel rod wall, and the tubes are sealed to contain the radioactive gases that are produced during the fission process.

The fuel rods are bundled into assemblies that enable the reactor technicians to load and remove many fuel rods simultaneously. The assemblies also hold the fuel rods in a configuration that permits the coolant to flow efficiently past them so that the thermal energy emitted by the rods is transferred to the coolant. Each assembly in the Westinghouse design is capable of holding up to 289 rods—some spaces are left empty so that they can later be loaded with other devices—and the assembly is inserted vertically into the pressure vessel. (Keep in mind that the details of the fuel rod and assembly designs vary according to the model of LWR considered. The figures given here are used to convey an idea of the scale of these devices.) In the Westinghouse pressurized water reactor considered in this paragraph a total of almost 51,000 fuel rods are loaded into the core on 193 assemblies.

Some of the empty spaces in the fuel assemblies described in the previous paragraph are filled with control rods. These are long, thin rods constructed of material that absorbs neutrons. One can

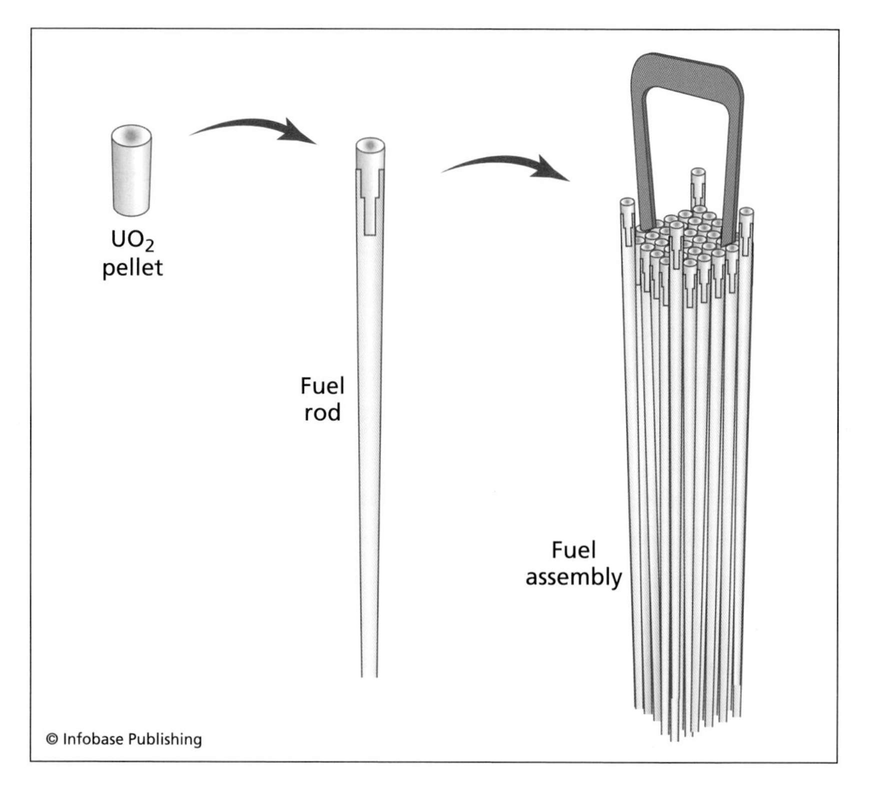

Fuel pellets are loaded into fuel rods, which are bundled into fuel assemblies. This configuration is of the type used in LWRs.

slow the rate of fissions—and so decrease the production of thermal energy—by inserting control rods into the assemblies. There are many control rods inside every LWR. The exact number depends on the particular design. Control rods allow operators to match the reactor power output to the electrical demand. By withdrawing the control rods one can increase the power output of the reactor. By fully inserting them one can bring the chain reaction to a stop.

CANDU reactor fuel and the fuel assemblies that contain the fuel differ from LWRs in several important ways. The fact, mentioned in the previous section, that they can operate with natural uranium is just one important difference. Another important difference between CANDU reactors and LWRs is in the construction of the fuel assemblies. As with the LWRs, CANDU fuel is formed from pellets of UO_2 that are inserted into zircaloy tubing. The difference is that there are far fewer tubes per assembly—28 or 37 depending on the assembly design—and the tubes are much shorter. CANDU fuel assemblies are only about 19 inches (48 cm) long. Each *fuel assembly* is inserted into one of many long horizontal pressure tubes. These tubes serve some of the same functions as the huge pressure vessels in the LWRs: They contain the fuel and the coolant. The coolant is heavy water and is maintained at about 100 atmospheres pressure to prevent it from boiling. (The heavy water moderator circulates outside the pressure tubes and is described in detail in the next chapter.) In a CANDU reactor, one can replace irradiated fuel with fresh fuel simply by switching out the small fuel assemblies even while the reactor is operating at full power. A fresh assembly is inserted into one end of the pressure tube, and a spent assembly is removed from the other end. This characteristic of CANDU reactors enables the reactor operators to exert more control over the way that fuel is consumed in the reactor. Replacing a few fuel assemblies is a procedure performed almost daily during the operation of a CANDU reactor.

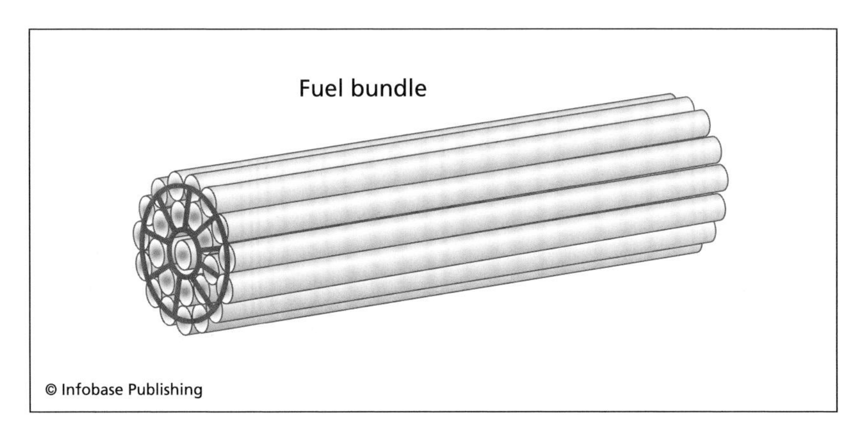

CANDU fuel bundles are much shorter than those of LWRs and can be inserted and removed while the reactor is in operation.

Once a new fuel assembly is inside the reactor, whether that reactor is a LWR or a CANDU reactor, the fission process is initiated and thermal energy begins to flow out of the fuel assemblies into the coolant surrounding them. The production of thermal energy is, after all, the purpose for which the fuel was created. Over time the fission process destroys an increasing percentage of the $^{235}_{92}U$ atoms, and in the process new and often radioactive elements are created. The newly created radioactive atoms begin the process of decaying into more stable isotopes, and this process also involves the release of thermal energy.

As the $^{235}_{92}U$ atoms are destroyed through the process of fission, some of the $^{238}_{92}U$ atoms undergo the process of transmutation described in the second chapter in equations (2.2) through (2.4), with the result that plutonium is created in the reactor fuel. Plutonium is fissile. A thermal neutron impinging on its nucleus will often produce a fission with the result that the plutonium nucleus splits to produce two new atoms and in the process a great deal of energy is released—an amount that is roughly equal to that produced when a $^{235}_{92}U$ atom is split. Reactors, therefore, not only consume fuel, they produce it. The twin process of fission and transmutation continue, and the products of fission accumulate in the fuel until their presence begins to interfere with the fission process, at which time the fuel is "spent" or exhausted. Spent fuel must be removed from the reactor. Unfortunately, what comes out of the reactor can be used for purposes other than the peaceful production of electricity.

NUCLEAR WEAPON PROLIFERATION

To create an atomic bomb—an explosive device that relies on the process of fission for its power—one needs fissile material. The most common fissile materials are $^{235}_{92}U$ and $^{239}_{94}PU$. (A more powerful type of nuclear weapon, the thermonuclear or hydrogen bomb, relies on the process of fusion for most of its power.) The bombs that

AVERAGE COMPOSITION OF SPENT BOILING WATER REACTOR FUEL			
ELEMENT	**GRAMS PER FUEL ASSEMBLY**	**ELEMENT**	**GRAMS PER FUEL ASSEMBLY**
Aluminum	31	Niobium	29
Americium	220	Nitrogen	25
Barium	390	Oxygen	25,000
Cadmium	27	Palladium	270
Carbon	36	Plutonium	1,500
Cerium	430	Praseodymium	200
Cesium	390	Rhodium	79
Chromium	1,900	Rubidium	64
Cobalt	26	Ruthenium	410
Europium	24	Samarium	160
Gadolinium	310	Silicon	80
Iodine	43	Strontium	110
Iron	5,100	Technetium	140
Krypton	62	Tellurium	91
Lanthanum	220	Tin	1,600
Manganese	160	Titanium	83
Molybdenum	630	Uranium	170,000
Neodymium	730	Xenon	950
Neptunium	97	Yttrium	81
Nickel	3,000	Zirconium	96,000

were exploded over Japan to end World War II were fission bombs. The bomb that exploded over Hiroshima, Japan, on August 6, 1945, used uranium with a high concentration of $^{235}_{92}U$, and the bomb that exploded over Nagasaki, Japan, on August 9, 1945, used $^{239}_{94}Pu$. One key objection to the use of nuclear power is that reactors produce $^{239}_{94}Pu$ as a byproduct. A second key objection is that the enrichment technology required to produce fuel for LWRs can, in theory, also be used to produce the material for nuclear weapons.

To be clear: No nuclear reactor of any design, no matter how it is operated, can cause a nuclear explosion. But uranium-235, a vital component of reactor fuel, as well as plutonium, a material found in spent fuel, can, when suitably processed, be used to produce nuclear weapons. The process of creating "weapons-grade" material is neither easy nor inexpensive, but it is well within the capabilities of many nations—the United States did, after all, create the first such weapons using 1940s technology—and the number of nations capable of producing such weapons is gradually increasing. Understanding some of the basic relationships that exist between nuclear weapons technology and nuclear reactor technology is crucial to appreciating the subject of nuclear energy.

There are two main areas of concern. First, as already mentioned, the technology used to slightly enrich uranium for use as reactor fuel can, without modification, be used to manufacture highly enriched uranium, also called weapons-grade uranium. Recall that the process of enrichment involves increasing the proportion of $^{235}_{92}U$ atoms to $^{238}_{92}U$ atoms in the fuel. The enrichment process considered in this chapter, which involves using centrifuges, enables the user to gain a very slight increase in $^{235}_{92}U$ at each step in the process. This process must be repeated many times to increase the percentage of $^{235}_{92}U$ in the uranium fuel from 0.7 percent to between 2 and 5 percent, the level of enrichment required for LWRs, but there is no reason why the operator must stop at a few percent enrichment. To obtain weapons-grade uranium, the process need only be repeated many more times in order to raise the percentage of $^{235}_{92}U$ to any desired level. In the centrifuge process one only needs many centrifuges, a large initial supply of natural uranium, a large energy supply—the process is energy-intensive—a high degree of technical skill, and patience. The same general ideas apply to other enrichment technologies. Uranium enriched to more than 90 percent $^{235}_{92}U$ is "weapons grade" material.

The second source of concern with respect to nuclear fuel involves the spent fuel. As the amount of $^{235}_{92}U$ decreases, the amount of plutonium in the fuel increases due to transmutation of $^{238}_{92}U$ into $^{239}_{94}Pu$, the other common material used to create atomic bombs. In the process of producing electricity, large commercial reactors produce quite a bit of plutonium. The exact amount of plutonium produced by a commercial reactor depends on the reactor design and the way that it is operated, but a reasonable estimate is that a large commercial nuclear reactor produces between 660 and 1,300 pounds (300–600 kg) of plutonium each year. This plutonium is not pure; it is mixed with many tons of spent fuel. Furthermore, separating the plutonium from the rest of the spent fuel is difficult, and when done improperly, the procedure is also dangerous. But the difficulties involved are not insurmountable, and a small nuclear weapon can be created from less than 10 pounds (5 kg) of plutonium-239. Different countries have different perceptions of the risks involved in *reprocessing* reactor fuel, which involves recovering the fissile and fertile material from spent fuel for reuse. France has a long history of safely reprocessing fuel, and one of the goals of their reprocessing efforts is to recover the plutonium present in spent fuel for use in a fuel called MOX, or mixed oxide fuel, a mixture of uranium and plutonium that can be used in nuclear reactors. By contrast, in the 1970s the United States decided not to reprocess spent fuel from commercial nuclear reactors. Officials acted out of concern about what would happen if the reprocessed plutonium-239 were to fall into the wrong hands.

Concerns about the relationships between enrichment and reprocessing technology and nuclear weapons proliferation regularly appear in the news. The long-standing controversy over Iranian plans to enrich uranium illustrates how these concerns are expressed in practice. As of this writing there is much in the press

The International Atomic Energy Agency

Established in 1957 and headquartered in Vienna, Austria, the International Atomic Energy Agency (IAEA) is an autonomous, intergovernmental organization composed of 2,200 professional and support staff. Its 2007 budget was approximately 283.6 million euros, which it received from member states. The IAEA has a broad mandate to develop and share expertise in peaceful applications of nuclear technology. It also monitors compliance with nuclear arms control agreements and develops safety standards for the nuclear industry. It has become increasingly involved in efforts to reduce the threat of nuclear-related terrorism.

Mohamed ElBaradei, director general of the International Atomic Energy Agency (IAEA)

The IAEA grew out of efforts by the United States during the 1950s to develop peaceful uses for atomic energy. To this end, the IAEA has fostered programs to use

about Iranian plans to enrich uranium, but the issue is hardly a new one. Iran began making covert efforts to acquire enrichment technology more than two decades ago, according to a report by Mohamed ElBaradei, the director general of the International Atomic Energy Agency (IAEA). The stated aim of the Iranian government

nuclear technology to improve agricultural crops. For example, 28 percent of Vietnam's rice exports consist of a strain of rice that is resistant to water with a high salt content, a strain that was developed in conjunction with the IAEA. Nuclear medicine is an important component in anticancer treatment therapies, and the IAEA works to improve access to nuclear medicine in the developing world. With respect to the problem of nuclear weapons proliferation, the main controlling document is the Treaty on the Non-Proliferation of Nuclear Weapons (NPT). With the notable exceptions of Israel, India, and Pakistan, all of which possess nuclear weapons, and Cuba, which does not possess such weapons, all other countries are signatories to the NPT, which seeks to foster peaceful applications of nuclear energy, prevent the spread of nuclear weapons, move all nations toward nuclear disarmament, and end the practice of testing nuclear weapons. To this end, approximately 900 nuclear facilities, including civilian and research reactors throughout the world, are subject to IAEA inspections to insure that those countries that are signatories to the treaty, first signed in 1968 and since extended, adhere to its conditions.

IAEA officials have played important roles in world affairs. Hans Blix, who was director general of the IAEA from 1981 to 1997, is especially notable in this regard. In 2000, he emerged from retirement to head the United Nations Monitoring, Verification, and Inspection Commission, which was responsible for weapons inspections in Iraq during the time preceding the second Iraq war. Mr. Blix famously resisted fierce pressure

(continues)

(continued)

and criticism from the United States to certify in the absence of proof that Iraq was in violation of weapons agreements that regulated its possession of "weapons of mass destruction." As noted in his March 7, 2003, report to the United Nations, delivered less than two weeks before the war, ". . . intelligence authorities have claimed that weapons of mass destruction are moved around Iraq by trucks. . . . No evidence of proscribed activities have so far been found."

As is now known, Mr. Blix's skepticism was well-justified. Iraq had no such weapons at the time of the war. The assertion that Iraq possessed such weapons was one of the main motivations for going to war.

Another equally notable IAEA official is Mohamed ElBaradei, who succeeded Mr. Blix as director general of the organization. In 2005 the Nobel Peace Prize was awarded to Mr. ElBaradei and the organization that he heads, the IAEA. The Nobel Committee praised Mr. ElBaradei's efforts to "prevent nuclear energy from being used for military purposes and to ensure that nuclear energy for peaceful purposes is used in the safest possible way."

is to manufacture slightly enriched uranium fuel for a domestic nuclear power industry. It is a controversial goal, but the issue of enrichment is not, according to the Iranians, open for negotiation. Iran claims a right as a sovereign nation to enrich uranium in order to manufacture nuclear fuel, and, in fact, as of this writing, the Iranian government is using the centrifuge procedure described earlier in this chapter to produce modest but increasing amounts of enriched uranium. The Iranian government has explicitly denied any interest in the acquisition of nuclear weapons. On November 15, 2004, Hassan Rowhani, secretary of Iran's Supreme National

Security Council, stated that, ". . . it [enrichment technology] is our legitimate right, and Iran cannot relinquish it." It is a position to which Iran continues to adhere.

Legally, Mr. Rowhani's assertion is not disputed by any nation. The Treaty on the Non-Proliferation of Nuclear Weapons, the main document governing the spread of nuclear weapons and the obligations of those countries already in possession of those weapons, does not deny the right of any signatory to the peaceful use of atomic energy. In fact, when the Iranian government temporarily suspended enrichment of uranium, the European Union explicitly recognized the Iranian action as "a voluntary confidence-building measure and not a legal obligation."

In response to Iranian assertions about its sovereign right to acquire and use enrichment technology, a number of countries, including those in the European Union, have expressed varying degrees of opposition to Iranian plans. Opposition stems largely from concerns about the possible consequences of Iran's possession and use of enrichment technology. Some countries are concerned that Iran's actual goal is to manufacture nuclear weapons. This is one reason why those countries that oppose Iran on the issue of enrichment technology express much less resistance to the idea of Iranian civilian nuclear reactors. In a counterproposal made in May 2006 on behalf of the European Union, Javier Solana, the foreign policy chief of the European Union, suggested that the European Union provide a state-of-the-art LWR to Iran, under the conditions that the fuel for the reactor be enriched by another country and that the Iranians agree to appropriate safeguards with respect to the disposition of the spent fuel. Iran refused.

The relationship between the peaceful production of nuclear-generated electricity and nuclear weapons can be complex, so it may help to put it in a historical perspective: No nation has ever used commercial nuclear power as a first step in the acquisition of nuclear weapons. All nations currently in possession of nuclear weapons—that is, the United States, Russia, France, the United Kingdom, China, India, Pakistan, North Korea, and Israel—developed the weapons first and a civilian nuclear power program second or not at all. (Israel, for example, has nuclear weapons and has never used its considerable know-how to produce any nuclear-generated electrical power for its citizens.) Additionally, despite the fact that a civilian nuclear power program *could* provide the basis for a nuclear weapons program, many nations have long possessed sophisticated civilian nuclear power programs and the necessary technical skills to develop nuclear weapons on short notice but have never done so. The inescapable fact is that nuclear weapons technology is now available to virtually any country with sufficient determination to develop it whether or not it possesses civilian nuclear reactors.

Nuclear Reactor Designs

Since the first nuclear reactors were created during the 1940s, a wide variety of creative designs have been proposed and many of those have been built—some large and some small. Historically, most reactors have used ordinary water as the coolant, but some have used heavy water—that is, water composed of oxygen and a rare isotope of hydrogen called deuterium—and still others have used helium or carbon dioxide. Historically, almost all reactors have used solid fuels, but a few research reactors have been built that used a liquid fuel. (The fissile material is mixed with a liquid in sufficient densities to sustain a controlled chain reaction.)

In addition to the larger, more familiar commercial reactors, many smaller reactors have been built, and continue to be built, for experimental purposes and for the military, where the reactors are usually used aboard ships. The five designs considered in this chapter are, therefore, not an exhaustive list. They were chosen for their commercial importance or because they demonstrate important principles.

When discussing the power output of reactors, it is important to keep in mind that nuclear reactor output is usually measured in two ways. The first measure of output is the amount of thermal energy produced by the reactor. The second way is to measure the amount of electricity produced by the reactor. To distinguish between the two, writers usually use a lower case letter—*t* for thermal and *e* for electrical—and so the thermal output of the reactor, measured in megawatts, or MW for short, is written MWt, and the electrical output of the reactor, also measured in megawatts, is notated as MWe.

BOILING WATER REACTORS

A simple schematic of a boiling water reactor (BWR) was used in chapter 1 to emphasize the similarities between this type of nuclear reactor and an 18th-century steam engine. A more complete schematic is used here, but one should not become distracted by the extra detail. The idea is still remarkably simple: Water is pumped into the reactor; it flows past the fuel assemblies where it serves to moderate the neutrons and absorb thermal energy from the fuel assemblies; some of the water is converted into steam; the steam is separated from the liquid water and, under pressure, is pumped toward the turbine; at the turbine the steam passes through a valve and is directed against the blades of the turbine causing the turbine to spin; once past the turbine, the steam is condensed and pumped back into the pressure vessel, where the cycle is repeated. This cycle characterizes all boiling water reactors.

To appreciate how this occurs in practice, consider Vermont Yankee, a boiling water reactor located in Vernon, Vermont. The plant is located near the Connecticut River on approximately 125 acres (50.6 hectares). The plant began operation in 1972.

As with all light water reactors, Vermont Yankee uses slightly enriched uranium as its fuel. The uranium is manufactured in the

Vermont Yankee nuclear power plant, a boiling water reactor *(Vermont Yankee)*

form of numerous cylindrical pellets of uranium dioxide (UO_2), each 0.5 inches (1.27 cm) high and 0.487 inches (1.24 cm) in diameter. The pellets are stacked into fuel rods made of a special zirconium alloy called zircaloy designed to be strong and radiation resistant and to conduct heat easily. The fuel rods are loaded into fuel assemblies. There are 49 fuel rods per assembly, and the reactor contains 368 fuel assemblies. This yields a total of 18,032 fuel rods containing 178,145 pounds (80,975 kg) of UO_2. Once in operation, the average surface temperature of a fuel rod is 558°F (292°C).

The fuel assemblies are located inside a large steel pressure vessel. The pressure vessel was manufactured in three parts: an

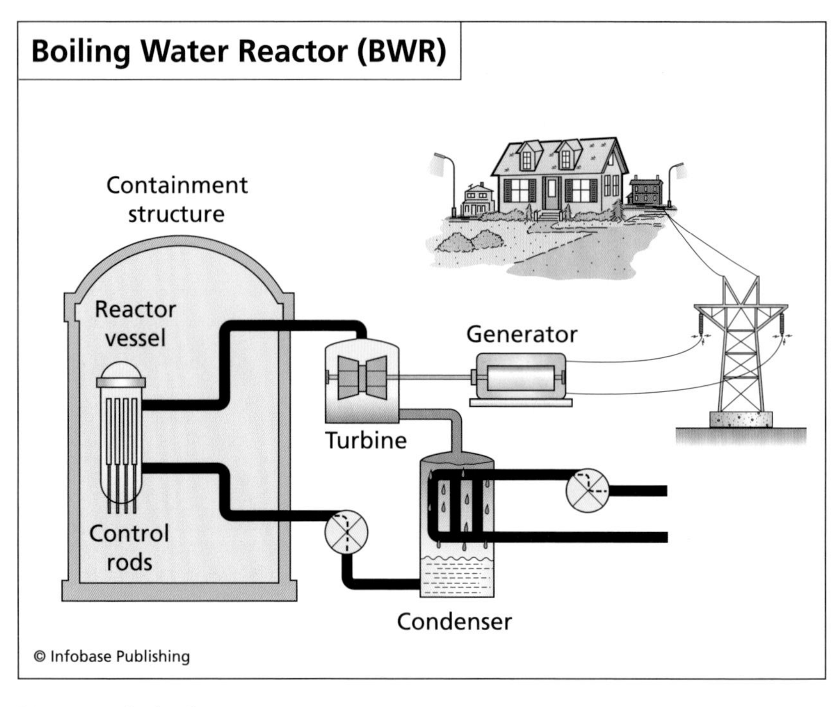

Diagram of a boiling water reactor

enormous cylinder, a "head" to cover the top, and a bottom. The pressure vessel was also manufactured with a steel lining made from a different alloy. The inside diameter of the cylinder is 17.1 feet (5.21 m), and the height of the pressure vessel is 63.1 feet (19.2 m). The cylinder, including the liner, is about five inches (13 cm) thick. Taken together, the cylinder, the top, and the bottom sections weigh 757,170 pounds (344,168 kg).

When in operation, coolant, in this case ordinary water, is supplied to the reactor at the rate of approximately 48 million pounds (22 million kg) per hour. The reactor produces steam at the rate of 6.43 million pounds (2.92 million kg) per hour.

The rate at which fission occurs is regulated, in part, with control rods, the long hollow steel rods containing neutron-absorbing materials. Vermont Yankee has 89 moveable control rods. The rods can be inserted into the pressure vessel or withdrawn from it. The deeper into the pressure vessel they are inserted, the more neutrons they absorb, causing the reaction rate to slow or even cease. When more power is required from the reactor, the control rods can be withdrawn, leaving more neutrons available for fission reactions, and in response the power output of the reactor increases.

Control rods wear out and must be replaced. The act of absorbing neutrons changes the rod's chemical composition. For example, boron-10, also written ${}^{10}_{5}$B, which is used in the control rods at Vermont Yankee, is changed when it absorbs a neutron in the following way:

$$ {}^{10}_{5}B + n \rightarrow {}^{7}_{3}Li + {}^{4}_{2}He $$

(The symbol ${}^{7}_{3}$Li is an isotope of the element lithium, and ${}^{4}_{2}$He denotes the helium nucleus.)

The control rods are a mechanical method of regulating the power output of the reactor, but a subtler method of control exists for boiling water reactors that is completely independent of operator intervention. It is built directly into the physics of the reactor and is called a "negative void coefficient." To see how this works, suppose that the control rods are partially withdrawn. This initiates the following sequence of events:

1. There is a brief surge in the output of thermal energy as the fission rate increases.
2. The water near the fuel rods, which moderates the neutrons as well as carries away the heat, begins to boil faster in response to the increased temperature of the fuel rods. As a consequence, more and bigger bubbles begin to form near the fuel rods.
3. Because steam is a poor moderator of neutrons, the harder the water boils, the fewer neutrons are moderated.

4. Unmoderated neutrons—that is to say, "fast" neutrons—rarely cause fissions. Therefore, as the number of moderated neutrons (also called thermal neutrons) decreases, the rate at which fission occurs also decreases, and the thermal output of the reactor decreases as well. All this occurs without operator intervention.

This is an example of a so-called *passive safety* feature. (It is sometimes also called inherent safety.) Passive safety does not depend on operator control or on mechanical devices. It is the best type of control because it is built into the operating characteristics of the reactor itself. By regulating the flow rate of the coolant supplied to the reactor, one can also use the negative void coefficient of the reactor to change the rate at which thermal energy is produced. The slower the water flows past the fuel, the more heat it absorbs, and the harder it boils. This is the beauty of this type of passive safety: Increases in reaction rates are automatically damped by the negative void coefficient of the reactor.

The pressure vessel is enclosed in a containment building, which is designed to remain intact even if the reactor vessel were to rupture, or if steam pipes were to break, and it is strong enough to retain water sufficient to flood the interior up to a level above the reactor core.

All of these systems and many more are designed to enable Vermont Yankee to produce 1,665 megawatts of thermal energy safely, about one-third of which, or 564 megawatts, are converted into electrical energy and transmitted to the consumer. The capacity of Vermont Yankee exceeds 90 percent—that is, it is producing its rated power more than 90 percent of the time. This is done on a comparatively small plot of land and with zero emissions. By comparison, producing the same amount of electricity from coal would result in venting more than 1 million tons of CO_2 into the atmosphere each year, a tremendous environmental burden.

PRESSURIZED WATER REACTORS

Pressurized water reactors (PWRs) constitute the most common type of commercial nuclear reactor. Some countries (France, for example) depend almost exclusively on PWRs. Other nations, the United States and Japan being two notable examples, use both BWRs and PWRs. The two types of nuclear reactors have a great deal in common: Both require enriched uranium for fuel; both use ordinary water—often called light water—to moderate neutrons and cool the reactor; and both employ massive pressure vessels, which are themselves situated inside even more massive containment buildings. Both generate steam to drive the turbine, and both have condensers whose function it is to change the steam, after it has driven the turbine, back into liquid water and so increase the efficiency of the process by which heat is converted into electricity. As the accompanying diagram indicates, they even look somewhat similar.

The most obvious difference between BWRs and PWRs lies in the way that steam is generated. Recall that in a BWR, the steam used to drive the turbine is generated inside the reactor pressure vessel at the surface of the fuel rods. By contrast, PWRs rely upon a device called a *steam generator.* PWRs are designed in such a way that very little

Oconee Reactor Complex. This complex of pressurized water reactors is one of the nation's largest producers of electric power. *(Ed Clem)*

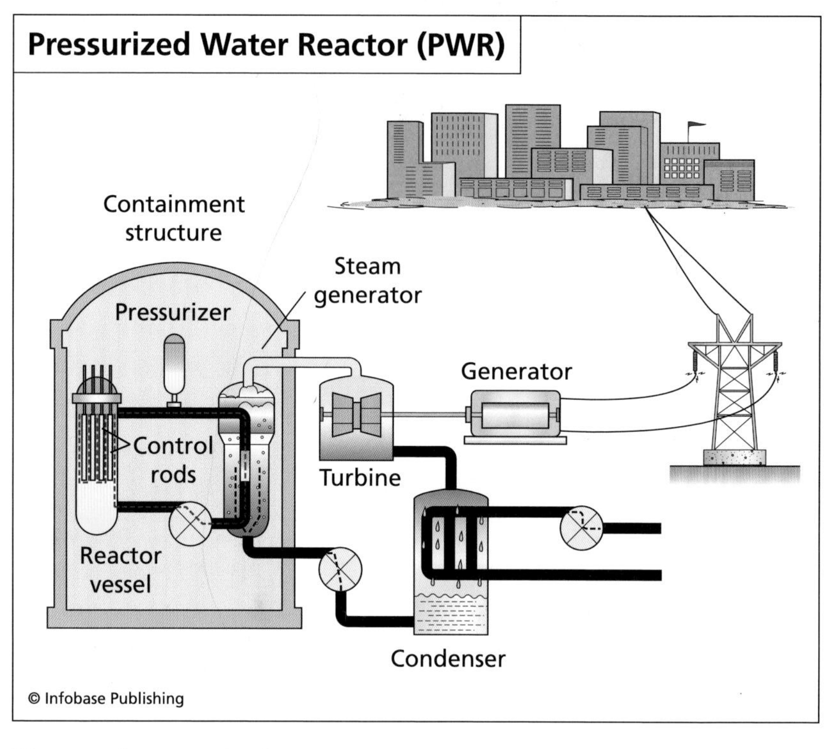

Diagram of a pressurized water reactor

steam is present in the pressure vessel. Instead, liquid water is pumped through the pressure vessel, past the fuel rods, where it is heated under sufficient pressure to prevent boiling. The heated water flows to the steam generator, where some of the thermal energy but none of the water from inside the reactor is transferred out of the reactor. The thermal energy that is transferred out of the vessel is used to generate steam in a secondary supply of water. It is this steam that drives the turbine. In a PWR, therefore, the steam supply and the coolant supply are physically separate. They circulate in separate "loops," or systems of pipes. The steam generators allow heat to flow between the two loops, while preventing them from exchanging water.

The decision to separate the process of generating heat from the process of generating steam in a PWR has other consequences. In particular, PWRs cannot depend on the negative void coefficient, described in the section on BWRs, to dampen power surges. Nevertheless, the safety record of PWRs is comparable to that of BWRs. Both have excellent safety records. The most serious accident at a Western nuclear reactor occurred in 1979 at a PWR at Three Mile Island, in Pennsylvania. The reactor was irreparably damaged, and the accident had far-reaching effects on the development of nuclear power in the United States, but no plant worker and no member of the general public was injured.

To appreciate the scale at which PWRs operate, consider the Oconee Nuclear Station located by Lake Keowee in Oconee County, South Carolina. The generating station is the site of three identical PWRs. As with Vermont Yankee, the fuel for the Oconee reactors is enriched uranium made into cylindrical pellets of uranium dioxide (UO_2). Different types of pellets may be used at Oconee, but they are all about 0.37 inches (0.94 cm) in diameter. They are loaded into fuel rods with a typical length of about 13 feet (3.94 m). The rods are made of an alloy called zircaloy-4 that is designed to be strong, radiation resistant, and have good heat-transfer properties to facilitate the movement of heat from the fuel to the coolant. The fuel rods are loaded into fuel assemblies. There are 208 fuel rods per assembly, and there are 177 fuel assemblies loaded into each core, for a total of 36,816 fuel rods. This means that there is, depending on the type of pellet used, between 92 and 98 tons (84,000–89,000 kg) of UO_2 fuel in each reactor. Each fuel rod has a full power lifetime—that is, at full power it is expected to last—480 days. (Not all of the fuel is replaced at once.)

The coolant flows past the assemblies, removing the thermal energy produced during the fission process. As it flows past the fuel assemblies at an average speed of about 15.5 feet per second (4.72 m/s), the coolant is heated to a temperature of 602°F (317°C) at a pressure of 150 atmospheres. The scale at which this occurs is

(continues on page 74)

The Accident at Chernobyl

On April 26, 1986, at 1:24 A.M., one of the reactors at the four-reactor Chernobyl nuclear power complex, which is located about 81 miles (130 km) north of Kiev, Ukraine, exploded twice. The first explosion, apparently the result of water flashing into steam, exposed the contents of the reactor to the outside air. (There was no containment building surrounding the reactor as there is for every Western reactor.) A few seconds later, a second explosion occurred when hydrogen and other materials from the reactor mixed with air and ignited in the heat generated by the reactor. The subsequent fires and the hot reactor core caused powerful updrafts that carried some radioactive materials high into the atmosphere, where they were spread by winds. How did this accident occur?

All four reactors at the site were of a Soviet-era design called RBMK-1000. At full power they produced 1,000 MWe apiece. Ordinary water was used as a coolant, and graphite, the same material used in Fermi's atomic pile, was used as a moderator. These reactors were designed with a positive void coefficient when operated at low power so that the production of steam within the coolant would cause an increase in the rate of fission. (Compare this with the negative void coefficient in Western BWRs.) Designers envisioned that the potentially unsafe positive void coefficient would be countered by strict safety procedures.

At the time of the explosion, the reactor's operators were performing an experiment at low power and had decided to ignore the safety protocols for the reactor. To perform the experiment, they completely withdrew the control rods from the core. The control rods, built in sections, were poorly designed: The upper part was designed to absorb neutrons, but the lower part was just graphite. When the control rods were fully withdrawn, there was a column of water beneath the graphite section. (Recall that ordinary water also absorbs neutrons.)

When, as part of the experiment, the operators decreased the volume of water moving through the core, the remaining water began to boil.

Because of the positive void coefficient, there was a surge of power as the fission rate increased. The operators attempted an emergency shutdown by reinserting the control rods. The descending control rods momentarily displaced the water beneath them with graphite. Because the water had been absorbing neutrons, the initial effect of the insertion was to further accelerate the fission rate and the production of thermal energy. Before the active part of each control rod was in place there was a tremendous spike in the thermal energy produced by the reactor. The water flashed into steam, and the reactor exploded. Less than one minute had elapsed between the time that the operators had begun the emergency shutdown procedure and the first explosion.

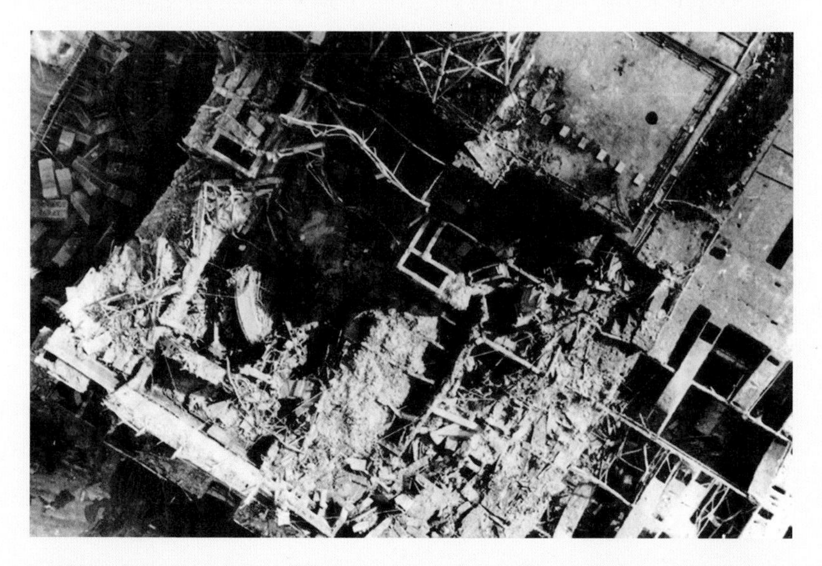

Looking into the debris-filled remains of the Chernobyl nuclear reactor *(GlobalSecurity.org)*

(continued from page 71)
hard to imagine: 65.66 million pounds (29.85 million kg) of water flows through each core every hour. Heat is transferred at the steam generators in sufficient quantities to produce 5.4 million pounds of steam per hour per steam generator at a temperature of approximately 590°F (310°C) and 62 atmospheres pressure.

Each pressure vessel at Oconee is about 40 feet 9 inches (12.4 m) tall and weighs 646,000 pounds (292,000 kg). The cylindrical shells have inner diameters of 14 feet (4.3 m). They are 8.44 inches (21.4 cm) thick.

Control of the reaction rate is maintained in several ways. There are two methods of particular relevance for this volume. First, each reactor uses control rods in a way that is similar to that of Vermont Yankee. The control rods are grouped into assemblies—16 control rods per assembly—and there are 61 control rod assemblies for a total of 976 control rods per reactor. Each assembly belongs to one of seven groups. Groups 1–4 are left out of the core and are maintained in that position unless it becomes necessary to shut down the reactor quickly. Groups 5–7 are inserted and withdrawn from the core as a way of actively maintaining control of the reactor's energy output. A second method of controlling the fission rate depends upon the use of boron mixed with the reactor coolant. As mentioned in the section on BWRs, boron absorbs neutrons. By adding boron to the coolant operators can slow the rate of fission, and by removing boron from the coolant operators can increase the rate at which fissions occur.

All of these technologies and many others enable the owners and operators of the Oconee nuclear station to maintain an output of 2,568 MWt per reactor for an electrical output of 846 MWe per reactor. As with Vermont Yankee and other old light water reactors, the reactors at Oconee convert about one-third of the thermal output of the reactor into electrical energy. Taken together, the three reactors comprising the Oconee station are, when measured by electrical output, one of the largest power-generating facilities of any type in the United States. A power source on this scale would be extremely challenging to replace.

CANDU REACTORS

An important alternative to light water reactors are the CANDU reactors. Canada has produced a variety of CANDU reactor designs since 1962. They share many features in common.

All CANDU reactors use heavy water as a moderator. Recall from chapter 3 that water molecules can be composed of oxygen and the most common isotope of hydrogen, which has a nucleus consisting of a single proton and no neutrons, or a water molecule may be composed of oxygen and deuterium, a rarer isotope of hydrogen that has a nucleus consisting of one proton and one neutron. Water that incorporates deuterium is often referred to as heavy water, because a unit volume of heavy water weighs about 10 percent more than the same volume of light water. Heavy water is an excellent

The Darlington Generating Station. This complex has four operating units, all of which use the CANDU design. *(Ontario Power Generation)*

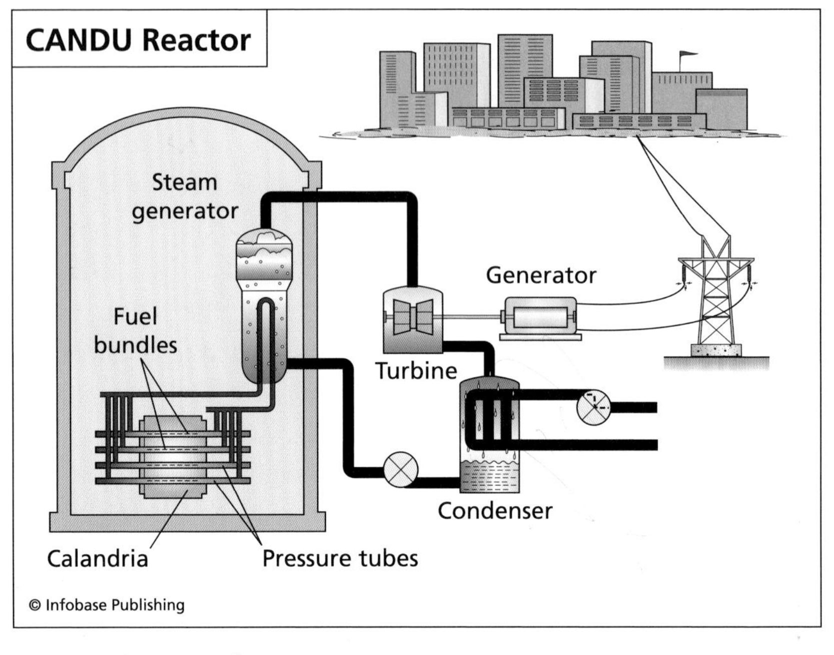

Diagram of a CANDU heavy water reactor

moderator of neutrons—much better, in fact, than ordinary light water, because heavy water absorbs far fewer neutrons.

Because they use neutrons more efficiently, CANDU reactors can use unenriched uranium as fuel. This is their usual fuel, but they also can use enriched uranium, MOX, or even fuel consisting primarily of the element thorium.

When Canada first began to contemplate manufacturing commercial nuclear reactors, engineers realized that it would be easier from an engineering standpoint to use a large collection of pressure tubes than a single enormous pressure vessel to contain the fuel. This accounts for the second remarkable characteristic of CANDU reactors. They have no massive pressure vessel. Instead, the reactor vessel, called a calandria, is a comparatively lightweight cylindrical structure that contains numerous horizontal tubes, called calandria tubes. Inside the calandria tubes are the pressure tubes. The space between the pressure tube and calandria tube is filled with carbon dioxide, and inside the pressure tubes is the fuel. CANDU fuel consists of small pellets loaded into fuel elements, short versions of the fuel rods in LWRs. Each fuel element is only about 19 inches (48 cm) long. They are bundled together into small fuel assemblies. Depending on the design, a fuel assembly might contain 28 or 37 elements—a 43-element bundle is under development—and these are loaded, one after another, into a pressure tube. The pressure tube is sealed, and pressurized heavy water is circulated throughout the tube as a coolant.

The pressure tubes and calandria tubes span the inside of the reactor vessel, which is flooded with heavy water. The heavy water that surrounds the calandria tubes functions as the moderator. Because the cooling system is so efficient at removing thermal energy, the moderator remains relatively cool and is maintained at a pressure of about one atmosphere while the reactor is operating.

From the point of view of safety, a ruptured pressure tube does not have nearly the same consequences as a ruptured pressure vessel in a LWR. Because each pressure tube is surrounded by a large source of relatively cool water, in the event that a pressure tube were to rupture the moderator could also act as coolant, and in any case a ruptured pressure tube exposes only a small fraction of the reactor fuel. By contrast, a ruptured pressure vessel in a LWR would expose the entire fuel supply. CANDU reactors are, in theory, highly stable.

The distribution of fuel assemblies into numerous pressure tubes also means that CANDU reactors can be refueled online—that is, in contrast to light water reactors, which are periodically shut down for refueling, there is no need to shut a CANDU reactor down to refuel. Indeed, some irradiated fuel assemblies are exchanged for fresh fuel assemblies almost daily in CANDU reactors. Online refueling is, in fact, one of the principal means by which reactor power output is controlled. The procedure is, in principle, very simple: A fresh assembly is inserted into one end of the tube and a spent assembly is removed from the other end. Online refueling means that, in theory, CANDU reactors have very high *capacity factor*s since there is no need to shut them down for refueling.

There are a variety of other means for controlling reactor output in a CANDU reactor, one of which is a system called liquid zone control compartments. The CANDU 6, a more modern CANDU design, has 14 tubes inserted vertically in the reactor vessel and positioned among the horizontal pressure tubes. These vertical tubes are further subdivided and partially filled with light water to absorb neutrons. By varying the amount of water in the tubes one can make small adjustments in the power output of the reactor. This is done to compensate for the small power variations that result from interchanging worn irradiated fuel assemblies with fresh assemblies.

Another method of controlling power levels, this time a method for increasing power in a CANDU 6 reactor, is a set of 21 so-called adjuster rods. These are made of stainless steel or cobalt and are normally fully inserted inside the reactor. They are withdrawn in order to increase reactivity. Mechanical control absorbers, which function in the same way that control rods function in light water reactors, are still another method of controlling reactor power output. (A separate set of control rods is used strictly for emergency shutdowns.) And still another method of control in a CANDU 6, a method similar to that used in PWRs, is a system for injecting a neutron-absorbing material into the heavy water moderator. CANDU 6 reactors use boron or the element gadolinium. (This is also an emergency control method: A second, entirely separate set of high pressure nozzles are used to inject a neutron-absorbing material into the moderator in the event an emergency shutdown is required.)

To convey the sort of scale at which large CANDU reactors operate, consider any one of the four nuclear reactors in Darlington, Ontario. These four essentially identical reactors are operated by Ontario Power Generation and borrow heavily on the design elements of the CANDU 6, but they are somewhat larger than the original CANDU 6 design. Each reactor is rated at 935 MWe. This four-reactor complex currently produces 17 percent of all the electrical power consumed in the province of Ontario. The calandria vessel for each reactor is 19.5 feet (5.98 m) long, and has an inner diameter of 27.75 feet (8.458 m), so it looks like a large squat cylinder lying on its side. Made from stainless steel, the shell is only 1.25 inches (3.17 cm) thick, and contains 11,019 cubic feet (312 m^3) of 99.95-percent-pure heavy water.

Each calandria shell contains 480 pressure tubes containing a total of 6,240 assemblies. The tubes have an inner diameter of only 4.07 inches (10.3 cm), and the thickness of the wall is only 0.16 inches (0.042 cm). The Darlington reactor uses a 37-element fuel bundle, each of which measures about 20 inches (50 cm) long and weighs about 53 pounds (24 kg). It contains about 42 pounds (19 kg) of uranium. Each reactor contains 237,600 pounds (108,000 kg) of uranium fuel.

The heavy water coolant, which flows through the pressure tubes at a rate of 55.4 pounds per second (25.2 kg/s), is heated to a temperature 590°F (310°C) at a pressure of 98.6 atmospheres.

Each calandria is enclosed in a containment building with walls that are six feet (1.8 m) thick. The Darlington reactors operate at about 32 percent efficiency, meaning they convert about one-third of the thermal energy produced by fission into electrical energy. All PWRs, BWRs, or CANDUs currently operate in the 30–40 percent thermal efficiency range.

Taken together, the various models of boiling water reactors, pressurized water reactors, and CANDU reactors account for most of the Western commercial nuclear reactors currently in service throughout the world. All three types of reactors have excellent safety records and produce large amounts of emissions-free electricity. Designs for all three types continue to evolve. By way of example, the so-called Advanced Boiling Water Reactor designed by

General Electric and a consortium of Japanese companies is a more sophisticated version of the boiling water reactor designs currently in use in the United States. The first advanced boiling water reactor began operation in Japan in 1996. Since that time, three more have been brought online in Japan, and additional advanced boiling water reactors are under construction in Japan and Taiwan. Evolutionary versions of pressurized water reactors—the new EPR built by AREVA NP has begun to be deployed—and improved versions of the CANDU reactors are also under development. (A new reactor is planned for the Darlington site.) The next generation of reactors, which are currently in the research stage, are called Generation IV reactors, and they will be very different from any reactor in operation today.

FUTURE REACTOR DESIGNS

One class of nuclear reactors in which there is a good deal of interest is called the pebble bed modular reactor, or PBMR. These reactors are designed to operate at much higher temperatures than today's designs. As indicated in chapter 1, a higher operating temperature also promises higher efficiencies. By contrast with the current crop of reactors, all of which convert about one-third of their thermal energy into electrical energy, PBMRs would convert almost half of the thermal energy they generate into electrical energy. They are also demonstrably more stable, and if brought into production, they would be cheaper and expose investors to less risk. The PBMR design, although considerably updated, is not new.

The German physicist Rudolph Schulten (1923–96) developed the concept of a pebble bed reactor in the 1950s, and a small reactor, built to test many of his ideas, began operation in the former West Germany in 1966. The reactor, called the AVR, operated successfully for about 21 years before it was shut down. The AVR used a so-called thorium-uranium cycle. The element thorium is more abundant than uranium, and while it has not been widely used as

Fuel sphere for a pebble bed modular reactor (PBMR)

a reactor fuel, it has the potential to substitute for uranium-238 in the sense that it functions as the fertile component of the reactor fuel: In the more common uranium-plutonium cycle, fertile $^{238}_{92}U$ is transmuted into fissile $^{239}_{94}Pu$; in the thorium-uranium cycle, fertile $^{232}_{90}Th$ is transmuted into fissile $^{233}_{92}U$. There is, however, no requirement to use thorium rather than uranium-238 in these reactors. The South African company Pebble Bed Modular Reactor Limited was established in 1999 to market these reactors.

Essentially all of the basic functions of a PBMR are performed in a way that is different from LWRs or CANDU reactors. First, the reactor's fuel is shaped into tiny particles, so small that they are difficult to see. The particles are coated by several layers of carbon compounds that seal the particle of fuel inside. The carbon compounds also serve to moderate the neutrons that are the result of fissions inside the fuel particle. A fuel particle and the multiple layers of material around it form a so-called microsphere less than one millimeter in diameter. The microsphere functions as a small pressure vessel, providing fuel, containment, and a neutron moderator. The microspheres are loaded into a spherical fuel pebble that measures

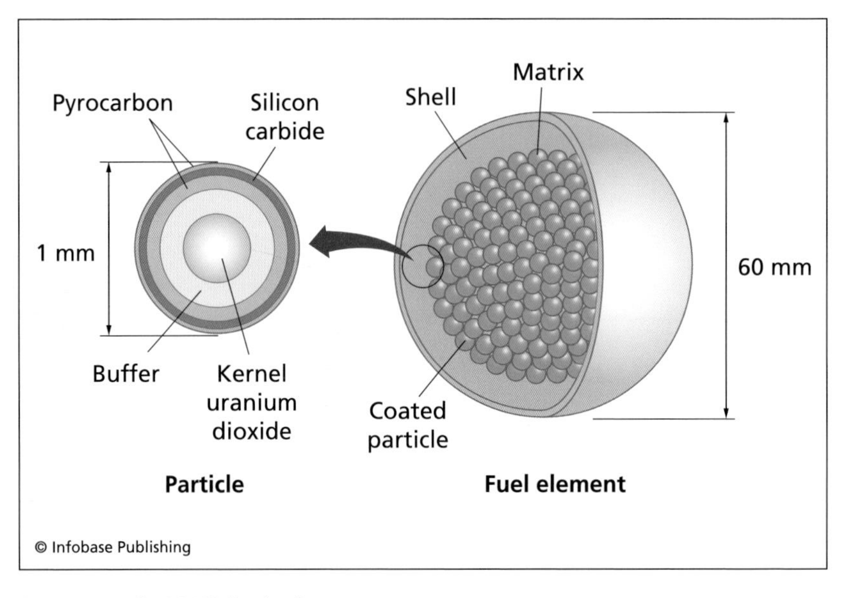

Structure of a PBMR fuel sphere

about 2.5 inches (6 cm) in diameter. Each fuel pebble contains about 11,000 microspheres packed into a graphite interior and covered with a protective layer. Hundreds of thousands of fuel pebbles are loaded—that is, they are dropped—into a reactor vessel, where the chain reaction is developed and heat is produced. Roughly 1 billion of the tiny microspheres would be contained inside a PBMR.

Second, the reactor requires no water. Helium is used as the coolant. The helium circulates through the spaces between the spherical fuel pebbles and acquires thermal energy in the process. It then flows out from the reactor to the turbine where some of its thermal energy is converted into work. The conversion of thermal energy into work results in a drop in temperature of the helium. Finally, the cooler helium is pumped back into the reactor vessel to repeat the cycle.

Each day fuel pebbles are removed from the bottom of the reactor vessel. Spent fuel pebbles are separated from the pebbles capable of contributing to the fission reaction that occurs in the reactor vessel, and these still-fresh fuel pebbles are dropped back through the top of the reactor vessel along with replacements for the spent pebbles.

These reactors are designed to be much more stable than the current generation of reactors. Their stability is, in part, the result of a phenomenon called "Doppler broadening." It works as follows: The hotter the uranium fuel becomes, the harder the molecules that compose the fuel will vibrate. As the temperature inside the reactor rises, the uranium-235 becomes less able to absorb free neutrons and so less able to produce additional fissions. Simultaneously, the uranium-238 in the fuel particles becomes more able to absorb neutrons. The result is fewer free neutrons. Consequently, the higher the temperature of the fuel, the more the fission rate drops. (The phenomenon of Doppler broadening occurs in every reactor, but the PBMR uses the effect more efficiently.)

Increasing the temperature of the fuel is less of a safety issue in a PBMR than it is in an LWR, because the reactor's fuel is far more heat-resistant. The carbon coating of the microspheres maintains its integrity at much higher temperatures than the fuel itself is capable of attaining. The coating remains undamaged at temperatures close to 3,600°F (2,000°C), which is well above the reactor's operating temperature. Since, even in the case of a severe accident, the maximum core temperature of a PBMR reactor would not exceed 2,900°F (1,600°C), these reactors can be operated safely without many of the active safety features upon which the current generation of reactors depend. In fact, to demonstrate the safety of the AVR reactor, operators in Germany stopped the flow of coolant to the reactor vessel and withdrew all control rods—that is, they maximized the reactor's thermal output while simultaneously minimizing the rate at which heat was removed from the reactor. The result? After an initial increase in temperature that did not get close to the temperature at which the fuel would be damaged, the power level of the reactor decreased and remained at a steady temperature indefinitely.

These reactors represent a new approach to harnessing nuclear energy. They may prove to be an important source of emission-free energy in the coming decades.

Another type of reactor that has attracted a great deal of attention from designers is the class of fast breeder reactors. These reactors produce more fuel than they consume, that is, they "breed" new fuel. To accomplish this feat, they depend on unmoderated or fast neutrons, which explains why they are described as fast. (In this section the terms fast breeder reactor, fast reactor, and breeder reactor will be used interchangeably.) The idea of a breeder reactor is not new. As early as the 1950s, scientists and engineers began to look for additional ways to extract more energy from nuclear fuel. It is easy to see why: Even today, in some countries—in the United States, for example—nuclear fuel is used only until the products of fission reach a sufficient density in the fuel to interfere with further fission reactions. At this point, the fuel has reached the end of its useful life, and the user disposes of the "waste." This is called a *once-through fuel cycle.* From an energy standpoint, a once-through fuel cycle is extremely wasteful because most of the energy in the reactor fuel has yet to be converted into heat. To make better use of reactor fuel, engineers can utilize the energy present in uranium-238, which is the main component of reactor fuel and which constitutes more than 99 percent of all global uranium supplies. (Fissile uranium-235 comprises less than 1 percent of all uranium on Earth.) Consequently, to make full use of uranium as an energy resource, engineers must find a way of fully "tapping" the energy in $^{238}_{92}U$.

Recall that although $^{238}_{92}U$ rarely undergoes fission, a $^{238}_{92}U$ atom will, under the right conditions, absorb an impinging neutron, and in the process be transmuted to plutonium-239 as described in equations (2.2) through (2.4) in chapter 2. As previously mentioned, plutonium is fissile, meaning it readily undergoes fission, and the fission of a plutonium atom releases roughly as much energy as the fission of a uranium atom. A fast reactor loaded with uranium fuel will produce power through the fission of uranium-235 and will simultaneously produce substantial amounts of plutonium through the transmutation of $^{238}_{92}U$. Some of this plutonium is consumed inside the reactor by the fission process, but much of the plutonium simply accumulates in the reactor fuel. The process does not continue indefinitely. Eventually, the products of fission accumulate to such an extent that they interfere with the fission process. At this point, the fuel is spent and must be withdrawn from the reactor. The same sort of process occurs in the commercial reactors in service today, but a fast reactor is designed to produce spent fuel that is unusually rich in plutonium. By employing reprocessing technology, the newly created plutonium and the uranium remaining in the spent fuel can be removed and combined to form a new type of fresh fuel, called mixed oxide fuel or MOX. The mixed oxide fuel can be reinserted into the reactor and used again. This process has the added advantage that the volume of waste produced by the reactor is a small fraction of that produced during a once-through fuel cycle.

Several fast breeder reactors have been constructed. The United States began operating a small experimental breeder reactor in 1951. It was the first nuclear reactor in the world to generate electricity, although it produced only enough for a few lightbulbs. The current leader in this field, however, is France, which has operated two sizeable breeder reactors, the 233-MWe Phénix reactor and the 1,200-MWe SuperPhénix. Fast breeder reactors have thus far been expensive to build and to operate, and they have not been as reliable as LWRs or CANDU reactors, but the technology is improving, and with fast reactors the world's uranium supplies can be extended far into the future.

Because they depend upon fast neutrons, fast reactors have a number of exotic-sounding properties. They have often, for example, been cooled by liquid metals. The reason is that the coolant must be chosen so that it does not also moderate the neutrons. Breeder reactor coolant should, therefore, satisfy the following three conditions:

1. It should have good heat-transfer properties; that is, the coolant should be efficient in conducting heat away from the core;
2. It should not absorb many neutrons; and
3. The coolant should not act as a moderator.

To satisfy the third condition, the coolant is often chosen so that the atoms of which it is composed are fairly massive. The situation is analogous to a billiard ball colliding with a stationary bowling ball. The billiard ball corresponds to the light quick neutron, and the bowling ball corresponds to the much more massive atoms that constitute the coolant. The greater mass of the bowling ball means that it will move only slightly in response to the force exerted by the billiard ball, with the result that the kinetic energy of the billiard ball is almost the same before and after the collision. In order to satisfy the three conditions required of a moderator in a fast reactor, engineers have generally relied upon liquid metals, especially sodium, although certain other materials could, in theory, work as well.

In addition to their ability to create their own fuel, fast reactors can, when operated in certain ways, also be used to destroy heavy metals found in spent fuel. The heavy metals, some of which would otherwise be radioactive for very long periods of time, are destroyed by neutron bombardment inside the reactor vessel, thereby producing materials that, though still radioactive, have much shorter half-lives.

Several new fast reactor designs are being proposed that would address the shortcomings of previous designs. These include reactors moderated with lead or sodium. Other fast reactors, using slightly different ideas, will depend upon a helium coolant. None of these reactors will be deployed for commercial service anytime soon. But the idea of building reactors that produce both fuel and electricity while simultaneously producing zero emissions is an idea that is too attractive for many to ignore.

Reactor Safety

The fission process is capable of releasing an enormous amount of energy per unit time. By way of example, it is estimated that roughly two pounds of highly enriched uranium underwent fission in the bomb that was detonated over Hiroshima. The rest of the mass of uranium was blown apart by the resulting explosion before it could participate in the chain reaction. Nuclear reactions are beyond our everyday experience both in terms of the amount of energy that can be released and the speed with which that release can be accomplished. Because of the tremendous amounts of energy released in these reactions, because of the levels of radioactivity produced by nuclear reactions, and because of the speed with which the reactions can take place, reactors are designed and operated according to very high safety standards.

How safe are Western reactors, the types of reactors now operating in the United States, France, and other developed nations?

Building one of the first next-generation pressurized water reactors, the European Pressurized Water Reactor, in Finland *(AREVA NP)*

Historically, the safety record is quite good. In the United States, for example, after thousands of reactor-years of experience, not one nuclear worker is known to have died from radiation exposure and not a single member of the general public has been exposed to a large dose of radiation. (A reactor-year is a measure of operating experience. One reactor operating for three years represents three reactor-years of experience, as do three reactors operating for one year apiece.) The safety records in other developed nations are comparable. Can one, therefore, conclude that Western reactors are safe enough to operate? Understanding the ideas involved in evaluating reactor safety requires an appreciation of engineering principles, mathematics, medicine, and other fields as well. The main danger posed by a reactor—and the only one examined in this chapter—is radioactive materials inside the reactor core escaping into the broader environment.

RADIATION EFFECTS

A large dose of radiation absorbed by the human body over a short period of time can lead to illness or death, but the effects that radiation has on human health depend on the type of radiation that is absorbed, the amount, and on the parts of the body that are exposed. The cells most vulnerable to radiation damage are cells that reproduce rapidly—skin cells, for example. Cancer cells, which reproduce rapidly and in an uncontrolled manner, are especially vulnerable to high levels of radiation, which explains the motivation behind treating cancer with radiation. None of this, of course, answers the question of how high a dose of radiation is dangerous.

Historically, radiation has been measured in a variety of ways. Different measures were developed for different applications. In attempting to describe the effects of radiation on human tissue, scientists have long recognized that damage is not a simple function of how much energy the tissue absorbs but also the type of energy. For a given energy, neutrons will, for example, produce more damage than X-rays. With this in mind, a particularly useful unit of measure is the *sievert* (Sv), which is designed to take into account the effects of different types of radiation. As a way of putting this into context, the Environmental Protection Agency (EPA) in studies of radon, a naturally occurring radioactive element present in the air of many homes, suggests taking remedial action if one receives a dose of about 0.008 Sv or eight millisieverts (mSv) per year due to the presence of radon in the air.

When a dose exceeding one Sv is absorbed over a brief time, the exposed individual will generally display symptoms of radiation sickness. Doses in the range of three to four Sv can be fatal. Symptoms include nausea, fever, abdominal pain, infection, and shock. Below one Sv, there are often no immediate symptoms. For lower doses of radiation—less, for example, than 100 mSv—there may never be any symptoms. The principal hazard associated with long-term exposure to low-level radiation is an increased risk of cancer.

The amount of increased risk depends on the degree of exposure. The type of cancer that might develop depends on the age of the individual at the time of exposure and the amount of time that has elapsed since the exposure took place.

The relationship between exposure to high doses of radiation and the resulting increased health risks is fairly well understood. The most complete data on the effects of large short-term doses of radiation come from long-term studies of the survivors of the World War II atomic bombings of the Japanese cities of Hiroshima and Nagasaki. The study group consisted of approximately 86,500 individuals. Of these, 50,000 received a dose of at least five mSv from the explosion. The remaining 36,500 individuals, who were generally located farther from the center of the blast, belonged to the control group, the group of unexposed individuals to which the exposed individuals were compared. Much of this research was carried out by the Radiation Effects Research Foundation (RERF). The accompanying table, taken from studies of the atomic bomb survivors done by the RERF, is an example of the way that different cancers respond to the same radiation exposure. In this case, research shows that leukemia is strongly correlated to exposure. Other data show that the closer the individual was to the center of the blast, the higher the leukemia risk. Children were especially susceptible, but in the case of leukemia, the risk was transient. The probability of developing leukemia peaked within 10 years after exposure and dropped steadily thereafter. Rates for several other cancers were shown to be somewhat elevated among survivors of the blast, but the effects were delayed by decades. For example, the data indicate that approximately 25 percent of all radiation-related cancers that occurred among the study group during the period 1950–90 occurred during the four-year period 1986–90.

The Life Span Study is extremely important because it is the best source of data on the relationships that exist between radiation exposure and health risks. Other studies—for example, research into relationships between radiation exposure levels as a result of the Chernobyl nuclear accident and the subsequent health of the exposed individuals—are also important. All of these studies help to identify the risks that radiation exposure poses.

Radiation doses and their relationships to health risks are often summarized in *dose-response curves*. There are many different dose-response curves. Each curve is an attempt to summarize in a simple, easy-to-grasp way the health risks associated with increased exposure to radiation. Mathematically speaking, a dose-response curve is a statement about a relationship that exists between the independent variable "dose" and the dependent variable "response," a term which is often, but not always, used synonymously for "cancer risk." Policy makers depend on these graphs when formulating safety standards and public policy with respect to nuclear power.

Details about the shape of dose-response curves vary depending on the specifics of the type of radiation exposure, the age of the af-

SUMMARY OF CANCER DEATHS IN THE LIFE SPAN STUDY COHORT OF ATOMIC BOMB SURVIVORS: 1950–1990

CAUSE OF DEATH	TOTAL NUMBER OF DEATHS	ESTIMATED DEATHS DUE TO RADIATION	PERCENT OF DEATHS DUE TO RADIATION
Leukemia	176	89	51
Other types of Cancer	4,687	339	7
Total	4,863	428	9

fected individuals, and numerous other factors, but there are commonalities. There is general agreement among researchers about the relationship between radiation exposure and cancer risk when the radiation exposure is fairly large—in excess of 100 mSv, for example. Most researchers would agree that the relationship between the two variables, dose and cancer rate, is linear, which means that the graph of a dose-response curve is, for doses in excess of 100 mSv, a straight line. To put this another way: A dose of $2x$ mSv of radiation, where x is greater than 100, produces double the response of a dose of x mSv of radiation.

The effects of very low doses of radiation—less, say, than 10 mSv—are less well understood, and this has important implications for those responsible for formulating public policy because it is exactly the effects of very low doses of radiation that are most relevant to the regulation of the nuclear power industry. The amount of radiation emitted by nuclear power plants is extremely low. It may well be lower even than the amount of radiation emitted by coal-burning plants. (Radioactive elements are present in coal in minute amounts, and because such large amounts of coal must be burned to produce substantial amounts of electricity, some of this radioactive material finds its way up the smokestack and out into the larger environment.) The EPA has set an exposure level of 0.25 mSv per year for members of the general pubic for radiation emitted by a nuclear plant. Actual exposure levels due to the operation of nuclear plants are far less than the 0.25 mSv standard. By way of comparison, a chest X-ray involves a dose of 0.7 mSv. None of these facts, of course, directly addresses the question of the seriousness of the health risks associated with low levels of radiation. The question is not easy to answer.

There are several reasons for the difficulties that scientists have encountered in attempting to precisely quantify the effects of low doses of radiation on human health. First, despite substantial research efforts, the biological mechanisms by which low doses of radiation affect health are less clear than is the case for higher levels

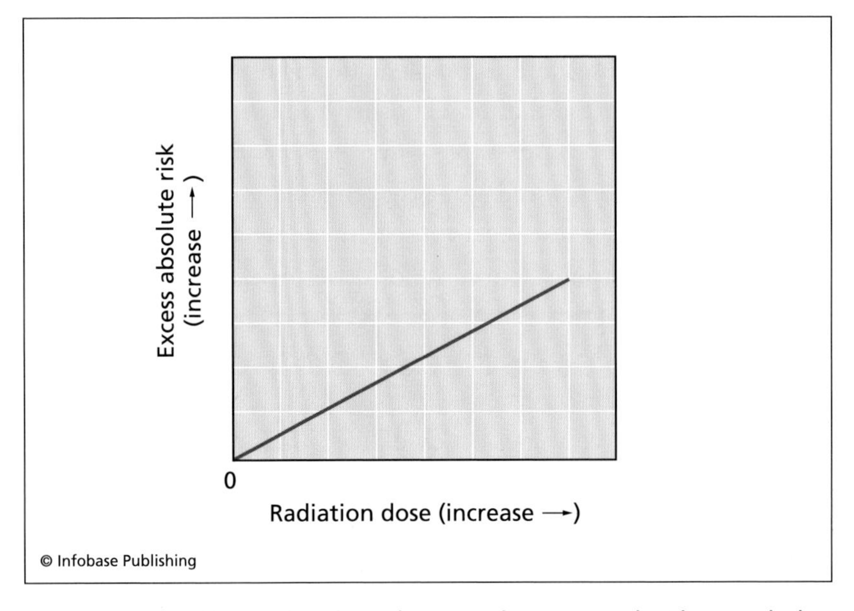

The no-threshold linear-dose hypothesis can be expressed with a graph that shows a straight line beginning at the origin and sloping upward as it moves right. (The steepness of the line depends on the units used along the axes.)

of radiation exposure. Second, everyone is already exposed and has always been exposed to low levels of so-called background radiation. The intensity of this naturally occurring radiation depends on where one lives and where one works. With respect to background radiation, most people absorb a few mSv per year, but levels of naturally occurring radiation can be higher. They are higher, for example, in parts of Colorado, where residents are exposed to levels of naturally occurring radiation that are two or three times the United States average. It is not clear that these increased doses lead to similarly increased responses. It is interesting to note that studies show that nuclear plant workers tend to be healthier than the general public.

For most citizens in developed countries, natural background radiation constitutes about 80 percent of one's radiation exposure. Much of the remainder is associated with emissions from the medical equipment and consumer devices that most of these individuals generally take for granted. Identifying the effects of any additional, relatively small amounts of radiation has proven to be very difficult. The problem is compounded by the fact that many people develop cancers that cannot be ascribed to any form of radiation exposure.

Currently, the most widely accepted approach to modeling the relationship between exposure to low levels of radiation (dose) and the risk of developing cancer (response) is to assume that the relationship is strictly linear. There is no evidence that assuming a strictly linear relationship understates the risk of developing cancer, and while there is some evidence that assuming a linear relationship may overstate the risk, the majority of those in the field currently prefer the more conservative assumption. This is the so-called *linear hypothesis.* In nonmathematical language, a linear relationship means that no matter what dose of radiation one is currently absorbing, if one doubles one's dose, one doubles one's risk of developing cancer. As a consequence, the EPA and other regulatory agencies have adopted the ALARA principle: restrict exposure to levels that are "as low as reasonably achievable." This is a flexible standard, because flexibility is necessary. Because radiation is a naturally occurring part of the environment, it is not possible to reduce radiation exposure to zero. And many believe that the benefits of certain other applications—diagnostic X-rays and large amounts of reasonably priced electricity with no CO_2 emissions, for example—justify the small additional risks associated with slightly elevated levels of radioactivity.

INDEPENDENT REDUNDANT SUBSYSTEMS

Nuclear power plant accidents that have the potential to endanger the public belong to one of two categories. The first type of accident with serious safety repercussions occurs when the energy output from the reactor increases in an uncontrolled way. This is called

(continues on page 98)

Where Are the Fatalities?

At 1:23 A.M. on April 26, 1986, a steam explosion, with an explosive power equal to 40 tons (36 metric tons) of TNT, blew apart the building housing the Chernobyl nuclear reactor. The explosion and subsequent fire lifted radioactive materials into the atmosphere. Much of this radioactive material settled in Ukraine, Belarus, and Russia. Lesser amounts settled elsewhere in Northern Europe; trace amounts were detected in North America. Chernobyl was by far the most serious reactor accident in history. What were its effects on the public health?

In September 2005 the International Atomic Energy Agency's Chernobyl Forum produced "Chernobyl's Legacy: Health, Environmental, and Socio-Economic Impacts," a report that identified two fatalities that occurred as an immediate result of the explosion: One body was not recovered and the other individual died of burns. In the months following the accident, 28 individuals, each of whom was either a first responder or a worker at the plant, died of acute radiation poisoning. In the years immediately after the accident there was a sharp increase in thyroid cancer among children as a result of drinking milk produced by contaminated cows. The cause of the thyroid cancer was radioactive iodine, $^{131}_{53}I$, which was deposited on grass; the grass was eaten by cows; the milk from the cows was distributed to local children; and the radioactive iodine in the cow's milk accumulated in the thyroid glands of the children. Thyroid cancer among children is usually extremely rare. Thyroid cancer among children was the principal measurable health effect of the Chernobyl accident. Four thousand cases occurred. This type of cancer is rarely fatal, but the treatment—which may involve surgical removal of the thyroid gland—adversely and permanently impacts those affected. Fifteen individuals had died of thyroid cancer by 2002. Measurements indicated that radiation exposure for most individuals in the general population, even among those living near the plant at the time of the accident, was low. Subsequent increases in most cancer rates, if they occurred, were too small to measure. In fact, no increase in the rate of leukemia, a disease known from studies of atomic bomb survivors to be caused by radiation exposure and to occur in the years immediately following exposure, has

been detected in the decades following the accident. The IAEA predicted a total of 4,000 to 6,000 deaths, including deaths that might arise decades after exposure, as a result of the accident. These would be concentrated among the 600,000 clean up workers, called liquidators, who worked in the area from 1986 to 1989, and a smaller number of evacuees, as well as individuals who remained behind. Researchers predict an increase of a few percent in the cancer rate for the individuals in this group.

These figures were disputed by Greenpeace, a nongovernmental organization, which issued its own report in 2006, "The Chernobyl Catastrophe: Consequences on Human Health." These authors predicted that upwards of 93,000 people would die as a result of radiation released from the Chernobyl reactor. What could account for such a huge difference in predictions made by the IAEA and Greenpeace, and should it not be possible to distinguish which prediction is more correct by examining cancer death rates?

(continues)

Chernobyl ghost town. A large region around the plant was evacuated after the accident. Much of it remains empty today. *(Elena Filatova)*

(continued)

The IAEA examined medical evidence and sought to draw conclusions from the available data. Because cancer rates among the general population could not—except for the very large increase in thyroid cancer among children—be associated with radiation exposure patterns, the IAEA concluded that any increases among the general population were too small to detect. By contrast, Greenpeace used the linear hypothesis to compute its figure. (Recall that the linear hypothesis asserts that if one doubles one's exposure to radiation, one doubles one's chance of developing cancer *and that this holds true no matter how small the radiation dose.*) Greenpeace examined the pattern of radioactive pollution that had formed across a large and heavily populated area of the world. Using the linear hypothesis, and the fact that radiation levels were slightly elevated over a very large area, Greenpeace computed a very small increase in cancer rates distributed throughout a huge population over a long time, an approach that results in a large number of predicted cancer deaths. But because so many people die of cancer anyway, the number of deaths attributable to Chernobyl would be so small a percentage of the total that it would be impossible to detect by examining public health data. It is an untestable prediction.

(continued from page 95)

a *criticality accident.* The second type of accident is a *loss-of-coolant accident,* which occurs when the cooling system fails to remove thermal energy from the reactor core at a sufficiently fast rate. Failure to remove thermal energy quickly enough can cause the fuel to overheat and even melt. Melted fuel has the potential to damage or even breach the reactor pressure vessel. Either type of accident, even if it is averted in its early stages, would generally mean an enormous financial penalty for the plant owner. From a public safety point of view—and this is the only issue considered here—the chief concern is that during the course of an accident some of the radioactive material inside the reactor vessel will find its way out into the surrounding environment.

Criticality accidents are extremely unlikely in all Western nuclear reactor designs. (Unless otherwise noted only Western reactors will be considered in this section.) The conditions under which a criticality accident might occur are well understood, and reactors are equipped with numerous safety systems to control the fission rate. Some systems are active—control rods, for example—and some are passive, systems that operate without any operator intervention. The negative void coefficient in boiling water reactors, described in chapter 4, is an example of a passive safety feature. Safety systems, whether passive or active, are designed to operate independently of one another so that the failure of one system does not cause the failure of another, and the simultaneous failure of multiple systems is extremely unlikely.

The other class of accidents that has potentially serious safety repercussions is the loss-of-coolant accident. These accidents continue to attract a great deal of attention from designers, regulators, and builders. The Three Mile Island nuclear plant accident, which occurred in 1979, is an example of a loss-of-coolant accident. (No one was injured in this accident, but the reactor was irreparably damaged.) One reason that loss-of-coolant accidents receive so much attention is that in the event that the cooling system, which is highly pressurized, is breached, a great deal of coolant can be lost very quickly leading to a rapid rise in fuel temperature.

To prevent the damage and potential dangers of a loss-of-coolant accident, reactors are built with multiple methods of supplying water to the core. These subsystems are redundant—that is, they have the same function—and they are designed to operate as independently of one another as possible so that the failure of one system does not interfere with the operation of another. This is the best way to ensure that in the event of a system failure, the function of the failed system is assumed by a backup system. The prospect that several independent systems would simultaneously fail is extremely unlikely. (Reactors are also built with sturdy containment systems to retain any coolant and steam that may leak from a break and to retain the contents of the pressure vessel should the pressure vessel fail.)

By way of example, in the BWR considered in chapter 4, the reactor is equipped with backup pumps. If the primary pumps were to fail mechanically, the backup pumps would take over until the system was shut down for repairs. If the primary pumps failed because of a power failure, a backup power supply would begin to provide power to the pumps until the system was shut down. If primary and backup pumps were to fail simultaneously, and/or if the power supplies (primary and backup) were to fail simultaneously, other procedures exist for keeping the core cool. In each case, the concept is the same. Each system is independent of its backup so that a particular function—in this case cooling—will fail to be performed only if all relevant systems fail simultaneously.

Another related approach to safety is illustrated by the many ways that the reactor fuel is kept isolated from the environment. First, the fuel, which is initially in the form of a fine powder, is formed into solid pellets, a process that inhibits dispersal of the fuel should it be exposed to the air. Second, the pellets are loaded into fuel rods, which are made of an extremely durable alloy called zircaloy. Third, the fuel rods, which form part of the fuel assembly, are sealed and placed inside the pressure vessel, an extremely strong steel chamber, which in the case of PWR is about eight inches (20 cm) thick. Fourth, the reactor pressure vessel is located inside a containment building, which is made of thick, reinforced concrete. All of these structures are independent of one another. The failure of one does not imply the failure of another, and the simultaneous failure of all of them is, therefore, extremely unlikely.

Passive and active safety features and a high level of redundancy all help to ensure the *safer* operation of a nuclear plant, but by themselves they do not answer the question: How safe are commercial nuclear reactors? To quantify the idea of safety, another set of ideas are required.

EVENT TREES

Risks—all types of risks—are measured in terms of probabilities. Every activity presents certain risks. Each decision to participate in an activity depends, in part, upon one's perception of the risks involved. Some risks are relatively easy to compute and are, therefore, well understood. The risks associated with driving an automobile, for example, are well understood. Driving occurs daily, as do the injuries and fatalities associated with driving. Actuaries and

Nuclear reactor control room *(NRC File Photo)*

others interested in the risks of driving continue to study driving statistics in the hope of refining their risk estimates, but the nature and magnitude of the risks associated with driving have, for the most part, already been computed. Risks associated with other activities—including the operation of nuclear reactors—are more difficult to quantitatively assess.

Evaluating the risks associated with nuclear power plant operation are more difficult than evaluating those for automobile operation because there have been so few serious accidents involving Western nuclear reactors. The most serious nuclear power plant accident to occur in the United States happened at the Three Mile Island reactor in 1979. The reactor vessel was irreparably damaged, but no one was injured. No member of the public has ever suffered a significant radiation dose from any U.S. commercial nuclear power plant during more than 3,000 reactor-years that the industry has been producing power. How does one evaluate the risk to the general public of nuclear power plant operation in the absence of significant accidents? One cannot conclude solely from the absence of injuries to the public that the technology is "safe enough to operate" without a rigorous definition of what that phrase means. The reason that a good safety record is necessary but by itself is insufficient to enable one to conclude that a particular type of reactor is safe is that, should a nuclear accident occur, the consequences could be extremely serious. Given the possible consequences of an accident, are the plants safe enough? Many of the concepts and techniques presently used to evaluate the safety of a plant and its subsystems belong to a branch of knowledge called probabilistic risk assessment (PRA).

Nuclear plants consist of well-defined systems, and each system is comprised of components that have been thoroughly tested. Even prior to the time that a component is installed in a plant, both operator and regulator have a good idea of the failure rate of the component in the sense that they can answer the following question: What is the probability that the given component will fail after *x* years of use? This question is answered by testing identical components under carefully controlled conditions. Each component, once installed, becomes part of a larger system. Plant operators maintain meticulous records about the reliability of each such system. They use this information to answer the following question: What is the probability that the system will fail after x years of use? Each system is part of an even larger functional unit, and again both operators and regulators use this information to establish the probabilities that each larger functional unit will fail.

But as with all complex machines, nuclear reactors are more than the sum of their component systems. Reactor systems must function as a whole. In the event that one system fails, the functions performed by the compromised system must be successfully passed to a secondary, or backup, system. Each sequence of events, together with the probability that the sequence will occur, is represented in what is called an "event tree." Each particular sequence of events is represented as a path through the event tree, in which each node identifies a junction where a particular function is passed from one system to the next. Each branch of a node is associated with the probability that the sequence of events includes that branch. By way of example, suppose that at a particular node one of two outcomes is possible—call them outcome A and outcome B. If $p(A)$ is the probability that event A occurs and $p(B)$ is the probability that event B occurs, then $p(A) = 1 - p(B)$ because there are only two outcomes at this juncture, A or B, and they are mutually exclusive. In other words, the sum of the two probabilities must equal 1. Each branch of the tree leads to the next node and so on until the path reaches its end, which represents the termination of the sequence of events. The probability that a particular sequence of events occurs is the product of the probabilities of the individual events in the sequence. Because nuclear reactors are so complex, their associated event trees are also very involved—too involved to show a complete tree on paper. Researchers have found that event trees enable them to separate the more likely sequences of events from the less likely ones. The more likely sequences are singled out for special study and remediation.

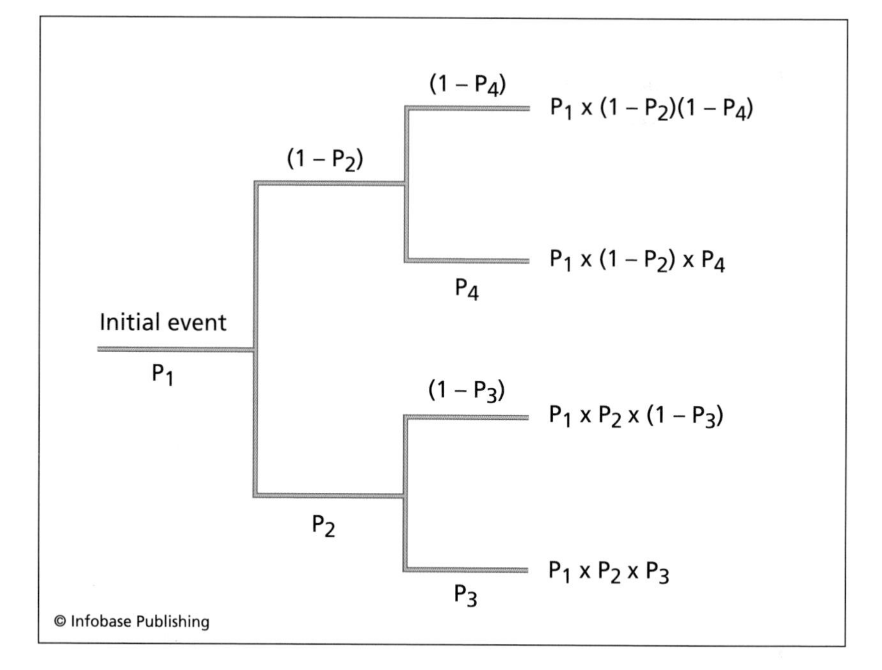

Model event tree. Each path from left to right represents a possible sequence of events together with the probability that the sequence will actually occur. (Event trees for real reactors are too complex to show on paper.) The probability of each sequence equals the product of the corresponding probabilities.

The first probabilistic risk assessment was performed in 1975 on a PWR called Surrey 1 located near Williamsburg, Virginia. The reactor was designed by Westinghouse, a leader in reactor design. The Reactor Safety Study, as it was called, made the assumption that the results obtained through its analysis of the Surrey 1 reactor were representative of all PWRs. Four years later, when the Three Mile Island accident occurred, the study was faulted for not identifying the sequence of events at Three Mile Island as a sequence that was comparatively likely to occur. In fact, the Reactor Safety Study estimated that the kind of damage experienced at Three Mile Island would only occur on average once every 100,000 years. But the reactor at Three Mile Island was a Babcock-Wilcox design, not a Westinghouse design. There were a number of important differences between the Westinghouse design, which had been analyzed in the Reactor Safety Study, and the Babcock-Wilcox design which was not analyzed. When the Reactor Safety Study methodology that had been applied to the Surrey 1 design was applied to the Three Mile Island design, it predicted that the sequence of events that actually occurred at Three Mile Island would occur on average once in 300 years—that is, the accident that actually occurred was roughly 300 times more likely to occur in a reactor of the Three Mile Island type than the Surrey 1 type. The lesson was learned. Now probability risk assessments are performed for every reactor.

Are reactors safe? The question cannot be answered on the basis of mathematical analyses and past practices alone, although it is irrational to ignore these factors. In the end, the question reduces to how much risk is predicted and how much risk one is willing to accept. Even then, the question must be placed in a wider context. There are few people in developed countries willing to live without electricity. Access to inexpensive and reliably produced electricity is a necessary precondition to modern life. In those regions of the world without electricity there are few people who do not want to enjoy the advantages of plentiful, reliable, and relatively inexpensive electrical power. The question, for those unwilling to live without electricity, must reduce to whether the risks and benefits associated with the operation of nuclear power plants are favorable relative to the risks and benefits incurred by generating the same amount of electricity through other means. The value of a particular power-generation technology should be measured in terms of its effects on the lives of the people who use it and on the environment in which they live. Effects, good and bad, are measurable. It is the responsibility of the citizens of each country to familiarize themselves with the available data and to base their decisions on the best information at hand.

Spent Fuel

The processes of fission, transmutation, and radioactive decay that occur as fuel is consumed within a nuclear reactor permanently change the chemical composition of the fuel. Spent fuel contains many elements that were not present in the fresh fuel. In fact, some materials found in spent fuel do not occur naturally on Earth. They exist on Earth solely because they were created within the reactor. Plutonium is but one example. Some of the radioactive isotopes present in spent fuel have very high levels of activity—that is, they are potentially very dangerous. Currently in the United States all of this fuel is stored at temporary sites. Most of these storage facilities are situated near the reactors where the radioactive waste was generated. This chapter describes the nature of spent fuel and some of the ways that governments have decided to recycle it—the technical term is *reprocess*—and/or dispose of it.

THE COMPOSITION AND CHARACTERISTICS OF SPENT FUEL

All radioactive elements begin to decay as soon as they have been created. One radioactive isotope decays to another, which decays to a third, and so on. The sequence ends when a stable isotope is produced. (A stable isotope is one that remains unchanged indefinitely.) This sequence of events is called a decay chain, and each isotope so produced can be pictured as a link in the chain. The half-lives of the isotopes formed in a decay chain may be measured in seconds, minutes, days, years, or millennia. Spent fuel is a complex and continually changing mixture of materials that generates heat and radioactivity for prolonged periods of time. It is important to understand, however, that as the various decay chains terminate in stable, nonradioactive isotopes, the amount of heat and radioactivity emitted by the spent fuel also diminishes.

By way of example, and using the notation developed in chapter 2, the accompanying chart shows the steps in the decay chain that begins with $^{238}_{92}U$. Because the steps are listed in the order in which they occur, each product element—that is, the element on the right side of each arrow—appears on the left side of the arrow in the next step of the chain because each product is the output of one step and the input of the next step in the sequence. (Recall that the term *half-life* denotes how much time elapses until half of the initial amount of a particular radioactive material remains.) In the following chart, the entry in the column labeled *half-life* refers to the half-life of the isotope appearing on the left side of the corresponding equation. Also, recall that the symbol *e* represents a beta particle, an electron ejected from the nucleus of the atom. When a beta particle is ejected from a nucleus, a neutron in that nucleus is transformed into a proton. The symbol $^{4}_{2}He$, identifies an alpha particle. As with the beta particle, the alpha particle appearing on the right side of an equation is ejected from the nucleus of the atom on the left side of the arrow. (Notice that the conservation laws listed in chapter 2 are satisfied at every step of the chain.) The chart is also remarkable because it illustrates the tremendous variation in the half-lives of some of the isotopes produced in a decay chain.

DECAY CHAIN FOR URANIUM-238

REACTION	NAME OF PRODUCT ELEMENT	HALF-LIFE
${}^{238}_{92}U \rightarrow {}^{234}_{90}Th + {}^{4}_{2}He$	Thorium-234	4.5 billion years
${}^{234}_{90}Th \rightarrow {}^{234}_{91}Pa + e$	Protactinium-234	24.5 days
${}^{234}_{91}Pa \rightarrow {}^{234}_{92}U + e$	Uranium-234	1.14 minutes
${}^{234}_{92}U \rightarrow {}^{230}_{90}Th + {}^{4}_{2}He$	Thorium-230	233,000 years
${}^{230}_{90}Th \rightarrow {}^{226}_{88}Ra + {}^{4}_{2}He$	Radium-226	83,000 years
${}^{226}_{88}Ra \rightarrow {}^{222}_{86}Rn + {}^{4}_{2}He$	Radon-222	1,590 years
${}^{222}_{86}Rn \rightarrow {}^{218}_{84}Po + {}^{4}_{2}He$	Polonium-218	3.83 days
${}^{218}_{84}Po \rightarrow {}^{214}_{82}Pb + {}^{4}_{2}He$	Lead-214	3.05 minutes
${}^{214}_{82}Pb \rightarrow {}^{214}_{83}Bi + e$	Bismuth-214	26.8 minutes
${}^{214}_{83}Bi \rightarrow {}^{214}_{84}Po + e$	Polonium-214	19.7 minutes
${}^{214}_{84}Po \rightarrow {}^{210}_{82}Pb + {}^{4}_{2}He$	Lead-210	0.00015 seconds
${}^{210}_{82}Pb \rightarrow {}^{210}_{83}Bi + e$	Bismuth-210	22 years
${}^{210}_{83}Bi \rightarrow {}^{210}_{84}Po + e$	Polonium-210	5 days
${}^{210}_{84}Po \rightarrow {}^{206}_{82}Pb + {}^{4}_{2}He$	Lead-206 (stable)	138 days

Uranium-238, the element at the head of this decay chain, decays very slowly, which is another way of saying that the amount of radiation it emits per unit time is very small. As a consequence, it has a very low level of toxicity and has been used in a number of applications—from specialized ammunition to ballast. Compared to uranium-238, many other materials found in spent fuel emit much more radiation per unit time. These materials also have much shorter half-lives.

MANAGING SPENT FUEL

Commercial nuclear reactors in the United States currently produce about 2,300 tons (2,100 metric tons) of nuclear waste each year. Since the early 1980s, the United States has planned to bury its spent fuel deep underground. France, by contrast, has long reprocessed its spent fuel in order to produce new fresh fuel. This illustrates the fact

Dry cask storage. Spent fuel, after it has cooled for years in pools of water, can be stored indefinitely above ground in these devices. *(Skoda J.S.)*

that, depending on the technology a country chooses to employ, spent fuel is either an energy source or a disposal hazard.

Nuclear waste from commercial nuclear reactors is solid waste, and in the United States there is a lot of it, and more is continually created. By the year 2030 the United States' reactors will have produced about 96,000 tons (87,700 metric tons) of spent reactor fuel. In addition to this spent fuel, there are also comparatively small amounts of nuclear waste from nuclear weapons research programs, nuclear weapons manufacturing operations, and research and military reactors. (Many Navy ships are powered by nuclear reactors.) About 90 percent of the waste currently destined for the yet-to-be-built nuclear waste repository at Yucca Mountain—about which more will be said later—is spent fuel from commercial reactors.

When fuel is removed from a U.S. reactor, it is left inside the fuel rods in which it was originally placed, and the fuel rods remain bound together in the fuel assembly. The goal is to dispose of the entire assembly as a single unit. Spent fuel is described in terms of its weight or its mass, but only the mass of the fuel itself is counted. Although current plans call for it to be buried along with the fuel, the assembly is not counted as part of the nuclear waste. The mass of the spent fuel is taken to be the mass of the fuel in the fuel rods prior to insertion into the reactor, because although the chemical composition of the fuel has undergone substantial changes within the reactor, the mass of the fuel has not.

The reactor fuel is, in any case, much more massive than the rods and assemblies used to enclose and arrange it. For example, prior to insertion into a PWR about 70 percent of the mass of the entire fuel assembly is uranium, 9 percent is oxygen—recall that the uranium fuel exists in the form of uranium dioxide or UO_2—and most of the remainder of the mass is the metal of the fuel rods and assembly structure itself. Roughly two-thirds of all high-level U.S. commercial nuclear waste is generated by PWRs and the remainder is generated by BWRs. In each case, once the fuel has served its function, the fuel assembly is taken from the reactor and placed in a pool of water to cool, because upon removal from the reactor the spent fuel generates a great deal of heat and radioactivity. Most of this energy is due to the decay of materials with short half-lives, so although it is initially intense it is not long-lived. During this time of high activity, immersion in water is necessary to keep the fuel cool and the plant workers safe.

After the first year, the amount of radioactivity emitted by the waste is about 1.3 percent of the intensity of the radioactivity emitted when it was first removed. After 10 years the fuel is generating radioactivity at less than 0.3 percent of the rate generated when it was first removed from the reactor. Despite the rapid decrease in energy emission, the fuel remains hazardous for a long time. Scientists and engineers have proposed a variety of ways to manage spent fuel, but broadly speaking all methods fall into one of two main categories: One can reprocess, or not. As mentioned at the outset of this chapter, some countries reprocess spent reactor fuel and others do not. Each country has what it believes to be compelling reasons for the method it has adopted, and it is instructive to review the ideas involved.

Currently, those countries that choose to reprocess spent reactor fuel do so with the primary goal of recovering the uranium and plutonium. The French, world leaders in the use of commercial nuclear reactors, reprocess nuclear fuel and are able to extract more than 99.9 percent of the uranium and 99.8 percent of the plutonium. (Plutonium can be mixed with uranium to produce a reactor fuel called MOX, or mixed-oxide fuel, and French reactors are now fueled with MOX.) There are two main advantages to reprocessing spent fuel: First, it greatly reduces the amount of radioactive waste requiring disposal. Second, it provides a source of reactor fuel that does not involve mining. (In Japan, for example, where there are few natural energy resources, reprocessing is viewed as a form of energy security.)

Reprocessing currently produces uranium, plutonium, and "waste." As already mentioned, countries that reprocess spent fuel use the plutonium and uranium, which constitute most of the reprocessed material, to manufacture new reactor fuel. A small amount of other long-lived fission products are also present within the spent fuel and these currently have little economic value. They can, however, be reinserted into specially designed fast reactors where they undergo transmutation by neutron bombardment and are destroyed. What remains is a comparatively small amount of highly radioactive waste consisting of radioactive isotopes with relatively short half-lives. One need only isolate these lighter isotopes for a few centuries (or less), after which time the amount of radioactivity emitted by them is very low. If "a few centuries" seems long, it is brief when compared with the million-year timescales envisioned by those countries that prefer to use nuclear waste repositories to isolate all of their spent fuel.

If reprocessing were without difficulties, however, every country with a nuclear energy program would, presumably, already have a reprocessing program in place. This is not the case. In particular, the United States has no reprocessing program, and its reasons for not establishing a reprocessing program reveal what some believe to be the central objection to the use of nuclear power to generate electricity.

Early in the history of the United States' nuclear program, the government envisioned reprocessing spent fuel. Even with 1960s technology it was possible to recover most of the uranium and plutonium. Beginning in 1966 the U.S. government operated a modest reprocessing facility at West Valley, New York. At the time, the advantages associated with reprocessing seemed clear to many, but by 1972 the West Valley facility had closed and eventually plans for other such facilities were abandoned. While it is true that at the time of the decision to close West Valley the price of uranium was so low that the plant was uneconomical to operate, the main objection to reprocessing fuel in the United States was not economic. It involved national security.

In 1977, the Carter administration abandoned all plans to reprocess nuclear fuel because of fears that terrorists might obtain the plutonium that is one of the main products of reprocessing. The reasoning is simple enough. Plutonium can be used to construct nuclear weapons—the first atomic bomb, tested in New Mexico on July 16, 1945, used plutonium, as did the bomb dropped on Nagasaki, Japan, on August 9, 1945—and all commercial nuclear reactors produce plutonium through the transmutation of $^{238}_{92}U$ as described in equations (2.2) through (2.4). The Carter administration decided that it was safer to leave the plutonium as a component of the highly radioactive spent fuel. (Separating the plutonium from the other materials in the spent fuel requires a significant technological investment. A low-tech approach to extract the plutonium would be ineffective and would almost certainly expose those who made the attempt to lethal doses of radiation.) The administration believed that the best way to protect the plutonium contained in the spent fuel was to leave the fuel as it was when it was removed from the reactor. The problem of producing plutonium as a by-product of reactor operation is a significant one: A single one-GWe reactor operating at full power for a year will produce approximately 740 pounds (330 kg) of plutonium, enough material to power about 40 nuclear weapons.

Plutonium is not the only material from which one can make an atomic bomb. Uranium can also be used, but atomic weapons that use uranium as an energy source require highly enriched uranium—that is, upwards of 90 percent of the uranium in an atomic weapon should be $^{235}_{92}U$ in order to create an explosion. By contrast, commercial reactor fuel is only slightly enriched. The proportion of $^{235}_{92}U$ in commercial fuel is roughly 3 percent—in natural uranium only about 0.7 percent of the atoms are $^{235}_{92}U$—and for this reason reactor-grade uranium cannot be used to make an atomic bomb. Moreover, the technologies available for enriching uranium are expensive, technically demanding, and require a substantial infrastructure. To be sure, uranium enrichment is a project that is within the reach of many governments, but as with reprocessing it is probably far beyond the reach of any terrorist group.

The question of whether spent fuel should be reprocessed was not one the United States alone could decide. The United States does not now, nor in 1977 did it have, a monopoly on the technology necessary to reprocess spent nuclear fuel. During the intervening decades, several countries have established reprocessing facilities and have reprocessed fuel for themselves and others with no security difficulties so far. It is not clear that the decision by the United States to forgo reprocessing enhanced its own security or the security of other countries. Nevertheless, the United States still has no reprocessing program in place, and it still plans to dispose of all spent reactor fuel in a specially designed waste repository.

Creating a repository requires both short- and long-term planning. Repository designers need to design a repository that is as resistant as possible to every conceivable method of sabotage—presumably a short-term goal with which most people now have some degree of familiarity—and the repository must also be able to safely retain its contents over very long time periods.

It is in designing a repository for the long term that a new type of thinking is required. In order to successfully build for the long term, engineers need to create a repository robust enough to withstand events so rare that they might only occur once in 10,000 years. Earthquakes, floods, and even the possibility of nearby volcanic eruptions need to be considered. The repository must be constructed in such a way as to minimize the likelihood of human intrusion even after the records of the repository's contents, purpose, and existence are lost, as must surely happen over such an enormous time scale. There is also the problem of predicting repository performance. Is it even possible to predict how the materials used in the construction of the repository might degrade during the thousands of years that the repository is supposed to function? Resolving these problems and others requires creative thinking and careful research.

BUILDING FOR ETERNITY

During the 1980s, having already decided against reprocessing its spent fuel, the United States began to search in earnest for a way of isolating nuclear waste from the biosphere, that part of the world occupied by living creatures, for an indefinite period of time. But the problem of isolating nuclear waste was not a new one. Beginning in the 1940s, large quantities of highly radioactive waste were generated by the United States as it sought to manufacture the first nuclear weapons. This material had to be isolated as it was produced. Serious discussions about the best way to manage highly radioactive wastes began in the United States during the 1950s as nuclear reactors, civilian and military, were brought online. Throughout the succeeding decades a wide variety of options were considered. In addition to geologic disposal, which involves placing the waste deep underground, other alternatives that have been considered involve placing the waste in deep ocean trenches, liquefying it and injecting it deep into the ground, and loading the waste on rockets and sending it into space. Today a few other nations have begun to plan for geologic disposal, but the United States began working on its geologic repository first. The work has not gone smoothly. (See chapter 8 for a description of the U.S. political experience in the construction of a geologic repository for nuclear wastes.)

Yucca Mountain, Nevada, a site located on government land near the place where the United States conducted aboveground and belowground nuclear weapons tests during the cold war, was chosen for evaluation and possible development. The process was begun in 1983 with the passage of the federal Nuclear Waste Policy Act and was significantly amended with the passage of the Nuclear Waste Policy Act of 1987. Revenue from a tax on energy use that is levied

Yucca Mountain, north and south portals *(Department of Energy, Yucca Mountain Project)*

on users of nuclear energy is placed into a fund. The fund is to be used to pay for all costs related to research, construction, operation, and closure of a future nuclear waste repository.

Yucca Mountain has three features that make it particularly suitable for a nuclear waste repository. First, the region around the site is sparsely populated. Second, there is little water in the area. A dry environment is important because water can seep through the ground and into the repository, where it can slowly dissolve the heavy metals in the waste. Given enough time, the dissolved radioactive materials can then migrate through the groundwater and make their way to the larger environment. Designers, therefore, wanted very much to prevent water from coming into contact with the waste. The Yucca Mountain site has been dry for many

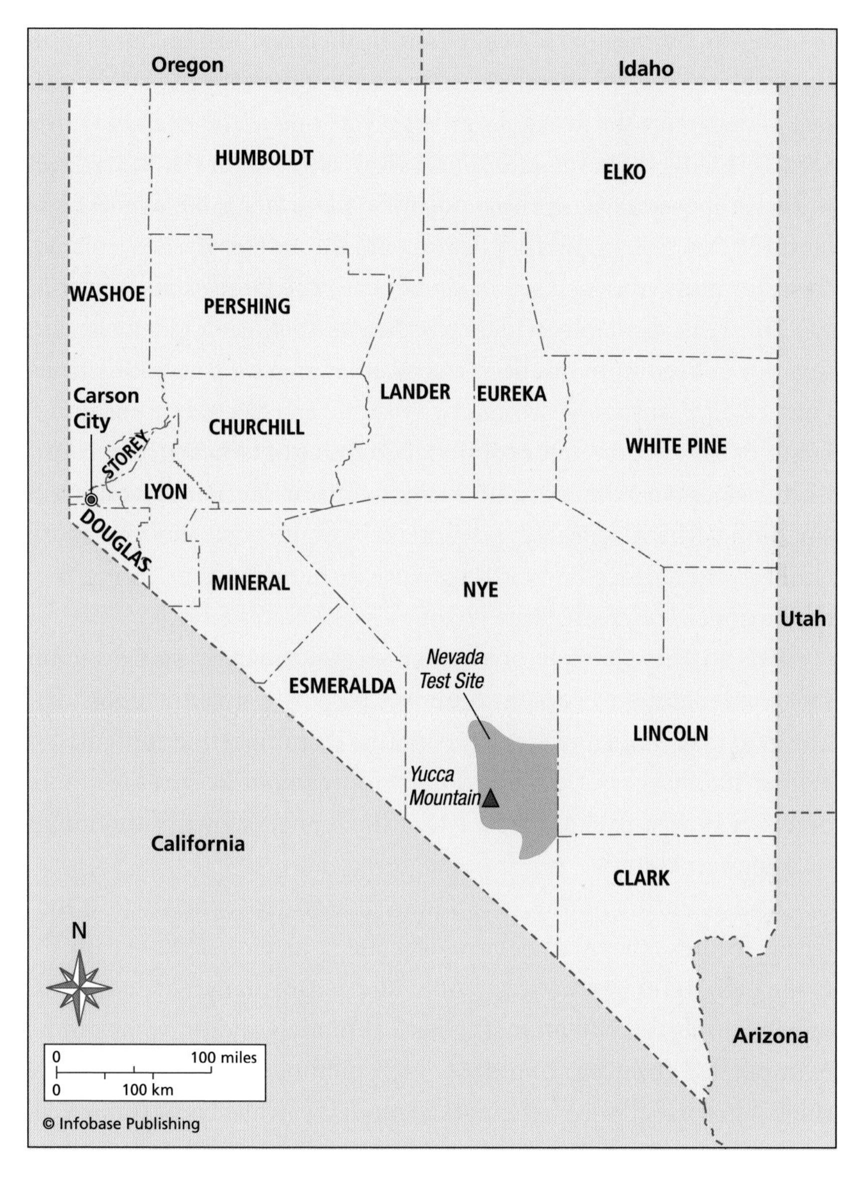

State of Nevada map, showing the location of Yucca Mountain Project

thousands of years, and there is currently very little precipitation in the area: The average annual rainfall is 7.6 inches (19 cm) per year. Virtually all of this water evaporates back into the atmosphere or is taken up by the sparse vegetation in the area. Just as important, the local aquifer, the water-saturated region beneath the land's surface that carries the area's water supply, is one of the deepest in the world—so low that even after burrowing down several hundred feet to the proposed level of the repository, the water table would be at least 530 feet (160 m) further down. And the aquifer itself is isolated. Its water flows into a closed basin beneath the Death Valley region.

Third, all available evidence indicates that, geologically speaking, Yucca Mountain has been a very quiet area for a very long time. Formed 12.8 million years ago by volcanic ash ejected by volcanoes north of the site, further eruptions are thought to be extremely unlikely. Moreover, there are no major faults at the Yucca Mountain site along which repeated earthquakes might raise the repository closer to the surface and in the process either expose its contents to the biosphere or create large changes in the water table.

It is in this sparsely populated, dry, seismically stable region that the federal government proposes placing a nuclear repository. The goal is to build a place that contains its radioactive contents for at least 10,000 years. Can it be done? How can one be sure—or more precisely how sure can one be—that the repository will function as its designers claim?

As a scientific matter, it is impossible to test construction materials for how they will behave over the course of 10,000 years. To make long-term predictions about the performance of materials, one must rely heavily on mathematical models and on what can be extrapolated from existing data. Such methods are helpful but are subject to correction. Mathematical models can always be improved, and later research may reveal new and important information. And the same shortcomings that call into question predictions by engineers about repository performance are even more pronounced in determining the reliability of the predictions of geologists. (The science of geology is more concerned with explanation than prediction.) One goal of repository designers is, therefore, to carry out their actions in such a way as to render them as reversible as possible. This enables later generations to benefit from advances in engineering and science by revising the work of previous generations and re-creating a repository more to their liking. They may even decide to remove the contents entirely. The Yucca Mountain repository is being designed so that its contents can be retrieved for at least 100 years, and possibly for as long as 300 years, after the first shipments begin to arrive. Should future generations decide to retrieve the contents for reasons of public safety, the protection of the environment, or to use the materials in the repository for their own purposes, they will be able to do so.

Envisioned as a large underground tunnel complex, the repository will, if it is built, be located at least 660 feet (200 m) below the surface. The tunnels of the repository are called drifts. Long, gently sloping ramps would connect the surface to the main drifts, and branching off from the main drifts would be many smaller chambers called emplacement drifts. The waste would be stored inside the emplacement drifts as indicated in the accompanying diagram.

The spent fuel would arrive at the site packaged in one of three types of shipping canisters. One type of canister can be placed directly in the emplacement drift upon arrival, but the other two types would require that the shipping canisters be further processed and their contents inserted into specially designed disposal canisters built to protect their contents for extremely long periods of time. Remote-controlled machines would transport the disposal canisters from the surface to the repository, where each would be placed inside a steel reinforced emplacement drift on top of a steel support structure called an invert and beneath a protective structure called a drip shield that would protect the canisters from any water that might make its way through the ceiling of the chamber.

One of the most important design questions involves choosing the so-called *thermal operating mode* of the repository. Because some of the energy radiating from the spent fuel manifests itself as

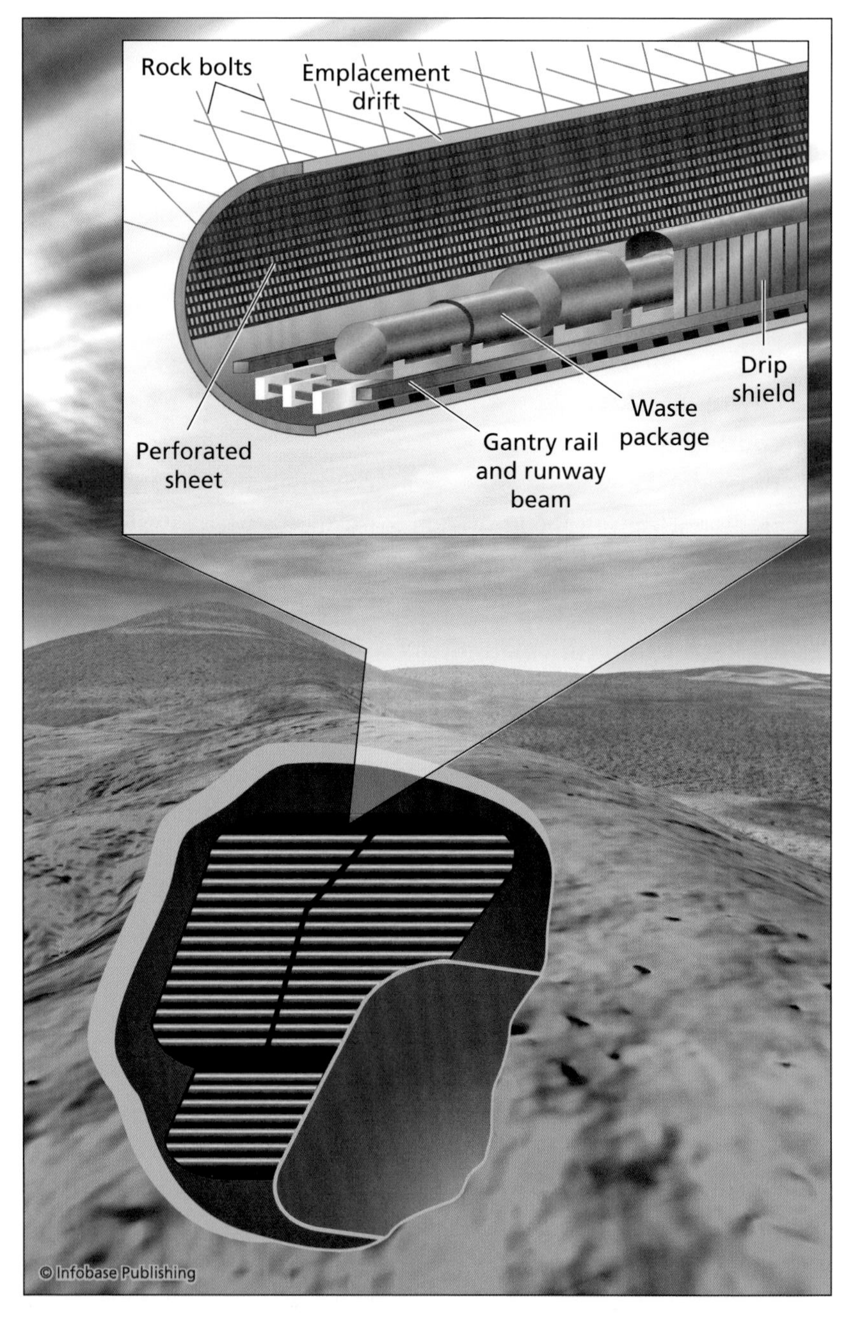

Cutaway view of proposed complex at Yucca Mountain and a cutaway view of an emplacement drift

heat, the waste packages continue to radiate substantial amounts of heat over many years as their contents undergo radioactive decay. If an emplacement drift were sealed shortly after the waste packages were placed inside—and here "shortly" means within 100 years—the temperature inside the unventilated drift would rise above the boiling point of water. Initially this was thought to be an advantage, because any water that percolated down from the surface would boil before it reached the emplacement drift. Upon further reflection, some called into question the effects of the extended high temperatures on the long-term performance of the waste packages themselves. More recently, engineers have considered a more conservative strategy of ventilating the complex for three centuries prior to permanent closure. A longer ventilation period keeps the temperature inside the emplacement drifts cooler, and during this time the amount of heat radiating from the waste packages would continue to decrease. Under this operating mode, the surface of the waste package would not exceed 185°F (85°C) even after the ventilation is turned off, the repository entrances sealed, the on-site buildings disassembled, and the repository closed forever.

How effectively will this scheme, as currently envisioned, contain the radioactive materials? The Department of Energy, using very conservative assumptions, estimates that over the time period that the site is open, no more than two cancer fatalities will result from the operation of the site. (No more than one cancer fatality if the repository is kept open for 100 years under the higher-temperature mode, and no more than two if the repository is kept open for 324 years under the lower-temperature mode.) Meanwhile, among those living in a 50-mile (80-km) radius of Yucca Mountain, the Department of Energy, using figures from the Centers for Disease Control and Prevention, estimates that during the operational period of the repository from 30,000 to 80,000 individuals will die from cancers completely unrelated to the operation of the repository. The repository represents, therefore, an increase in the local cancer rate of no more than 0.007 percent.

The repository was, of course, initially designed to isolate its contents from the environment as thoroughly as possible for a period of not less than 10,000 years, so at some point, barring a decision by future generations to remove the contents, the repository must be permanently sealed; the buildings that had been constructed to receive and process the waste will be dissembled, and the area will be returned as nearly as possible to its natural state. Once the environment has been restored to its original condition, this enormous project will have been completed. How well is it expected to work?

The Department of Energy's *Final Environmental Impact Statement for a Geologic Repository for the Disposal of Spent Nuclear Fuel and High-Level Radioactive Waste at Yucca Mountain, Nye County, Nevada,* contains the estimate that "Under the entire range of repository operating modes, less than 1 latent cancer fatality would be likely over the 10,000-year analysis period." This is an extraordinary claim. Keep in mind, however, that the waste will be separated from the surface by several hundred feet of rock. Simultaneously it will be located hundreds of feet above the water table and the fuel will be enclosed inside multiple engineered barriers specially designed to withstand the test of time. The Yucca Mountain repository is one of the most ambitious, most reviewed engineering projects ever attempted.

IS YUCCA MOUNTAIN SAFE ENOUGH?

Yucca Mountain may or may not prove to be the final resting place for a great deal of toxic material currently stored in pools of water on the grounds of nuclear power plants. It is, however, beyond dispute that spent fuel is currently less protected than it would be if it were buried beneath almost 700 feet (213.4 m) of rock. In that sense, Yucca Mountain represents a clear improvement over the status quo. But how safe would this waste be if it were buried at Yucca Mountain?

This device bores through the mountain leaving a tunnel in its wake. The Yucca Mountain project is one of the most ambitious engineering projects in history. *(Department of Energy, Yucca Mountain Project)*

Questions about safety are more than technical questions; they are questions about values. Answers depend, in part, on the amount of risk that one is willing to accept—in this case health and environmental risks associated with building or not building the facility at Yucca Mountain—and answers also depend on the value that one places on a particular benefit, in this case large amounts of moderately priced emissions-free electrical power. The situation is further clouded because the dependability of 10,000-year predictions is open to question. Various "precise" numbers are given for a number of safety parameters, but the meaning of the numbers is not always clear. Consider the conclusions of the Environmental Protection Agency (EPA) in establishing the permissible amount of radiation that can escape from the repository in the distant future:

The EPA established a radiation dose limit of 0.15 millisieverts per year (mSv/yr) for any member of the general public from the

Yucca Mountain facility. This is an extremely low limit, and it is currently not clear what the relationship is between cancer rates and such a low level of exposure. By way of comparison, it is estimated that *in their own homes* Americans are exposed to much higher levels of radiation due to radon in the air and drinking water than the standard imposed by the EPA at Yucca Mountain. Exposure levels depend on location, but in Connecticut, for example, the average exposure due to radon in the air of residential basements is about 3.5 mSv/yr, which is roughly 23 times the maximum of 0.15 mSv/yr established for Yucca Mountain. The EPA has what it calls an "action level" for residential radon levels that is roughly 40 times higher than the level established for Yucca Mountain. Such huge discrepancies between the safety standards established for people alive today and for those yet-to-be-born individuals, who may or may not live near Yucca Mountain, are difficult to justify on any scientific basis.

Another possible objection to the Yucca Mountain repository arises on the basis that 10,000 years is too brief a time for which to plan. The DOE's own calculations indicate that maximum release rates of radiation, while still quite low, would occur a few hundred thousand years after the repository is closed. Operating under the assumption that these calculations are reliable, some prefer to see the government predict the release rates and safety of the repository for much longer periods—up to a million years—and to design accordingly. These parties have found support in a 2004 judicial decision that attempts to require the federal government to substantially broaden the time period under consideration. Whether it is currently possible to make reliable forecasts over such enormous times is open to question, as is engineers' ability to design with such forecasts in mind.

Alternatively, some argue that 10,000 years is already a very long time—too long to reliably predict whether the repository will function as predicted. Over how long a period can one reliably predict

Nonscientific Objections to the Yucca Mountain Site

"Congress apologizes on behalf of the nation to the individuals described in subsection (a) for the burdens they have borne for the nation as a whole."

—*from the 1990 Radiation Exposure Compensation Act*

There are purely unscientific reasons to object to the location of the repository. Among many people there is a deep distrust of the federal government on matters relating to nuclear energy, particularly in Nevada and Utah. The aboveground testing of atomic weapons at the Nevada Test Site from 1951 until 1962, even given the security concerns of the time, can reasonably be described as callous. Atomic weapons were detonated aboveground only when the wind blew away from metropolitan areas and toward more sparsely populated areas in Utah and Nevada, a policy that caused illness and death among some inhabitants living downwind of the tests. Even if one concedes the necessity of testing weapons in that way at that time, there can be no excuse for not warning those who would be affected by the resulting radioactive fallout. But an intentional failure to warn is exactly what happened. Such behavior is certain to engender justifiable suspicion of future government assertions about the safety of any nuclear endeavors.

It has long been recognized that cancer rates were elevated in these regions in the years following the tests as a result of the detonation of atomic weapons, but it was only in 1990, almost 30 years after aboveground testing ceased, that the federal government offered an apology, one sentence from which is quoted at the beginning of this section, and modest compensation to certain individuals or their survivors for the illnesses that arose as a result of their exposure to radioactive fallout.

There are also some who oppose any technology that would facilitate the permanent and safe disposal of nuclear wastes. For these parties, disputes about nuclear waste substitute for what they believe to be the real

(continues)

(continued)

issue: stopping the further development of nuclear power. By blocking any solution to the question of permanent waste disposal, these parties hope to end nuclear power as an energy option. Without a place to safely dispose of the waste, it would be unwise to continue to generate it. Consequently, as each so-called safety objection is satisfied, a new objection is raised for the purpose of impeding the development of nuclear power.

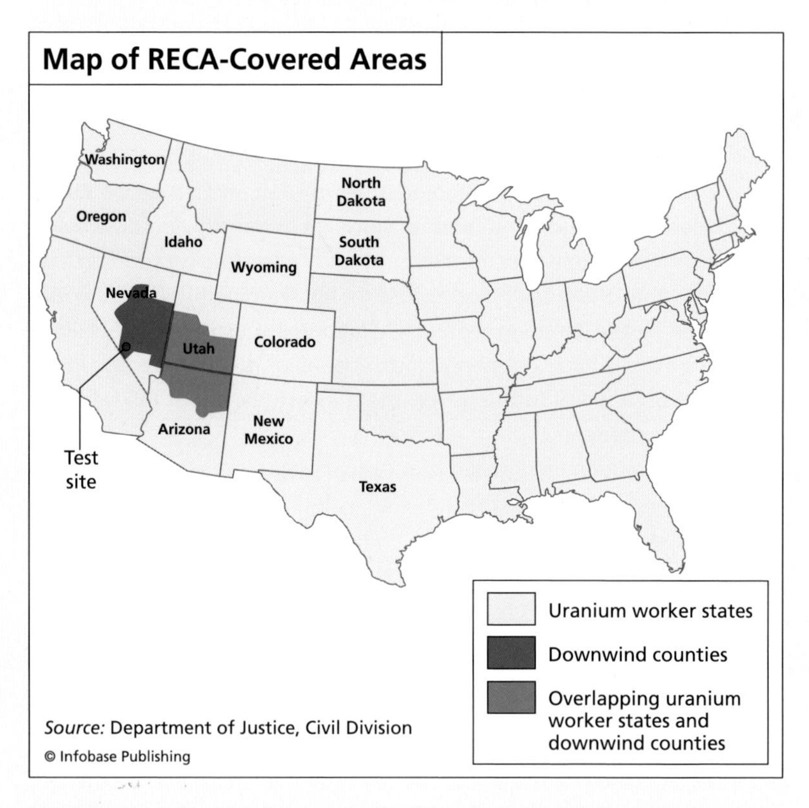

Map showing the area "downwind" of the aboveground nuclear weapons test site *(RECA stands for Radiation Exposure Compensation Act)*

geologic and climatic conditions? Is it reasonable to attempt to predict how the newly created alloys used in the construction of the waste packages will behave over 10,000 years? How can one test such predictions? These questions are also extremely difficult to answer.

The commercial nuclear power industry in the United States is about 50 years old, and the government still has no coherent policy for managing spent fuel. It neither reprocesses spent fuel nor does it bury it safely out of harm's way. After more than two decades of research and construction not a single fuel rod has been buried at Yucca Mountain. All indications are that it will be at least another 20 years before an underground repository begins operation. For the United States all of the hard decisions lie ahead.

The Business of Electricity Production

Nuclear reactors occupy a specific niche in the power sector. They can supply, without interruption, large amounts of power for long periods of time, but they are poorly suited for quick start-ups and shutdowns. Taken together, these two facts confine nuclear power plants to a narrow but very important role in the power generation sector. To understand the way that nuclear power is used, it is necessary to understand something about the way that demand for electricity fluctuates as well as the structure of the electricity business. These are the goals of this chapter.

UTILITIES AS NATURAL MONOPOLIES

In 1879, following numerous unsuccessful attempts at his laboratory in Menlo Park, New Jersey, the American inventor Thomas Edison (1847–1931) finally made a working incandescent lightbulb. The bulb burned for 44 consecutive hours before failing. This pro-

Power lines. An enormous demand requires an enormous supply. *(Ray Montoya, Sandia National Laboratories)*

totype proved the practicality of the idea. All that remained was to bring his bulb to market. This proved to be more difficult than one might imagine.

Edison's efforts to capitalize on one of his most famous inventions were hampered not only by a lack of durable lightbulbs and lightbulb manufacturing facilities but also by a lack of infrastructure to provide the electricity upon which his bulbs depended. As a consequence, Edison had to do more than manufacture improved versions of his lightbulb. He required a system to generate electricity and a system to transmit electricity to the lightbulb user. The generators, the transmission system, and the auxiliary hardware that grew out of his efforts were all wholly owned assets of his electric company. For succeeding generations, Edison's company was the prototype of a successful electric company: a local monopoly that controlled all aspects of electricity generation, transmission, and distribution—it controlled everything, in fact, except the lightbulb.

During the decades after the establishment of the first electric companies, the number of small and large power companies grew quickly. Despite differences in size and complexity, most of these companies shared a common structure: They were stand-alone entities, each responsible for all aspects of electricity generation, transmission, and distribution within their service areas.

Unregulated monopolies in all sectors of industry have a well-documented history of raising prices while simultaneously cutting back on service, and there have been times when, for the public good, the federal government has broken up monopolies. But for many decades this did not happen in the electricity industry because the electricity business was perceived as a natural monopoly. Supply, distribution, and transmission were all perceived as different aspects of the same function, and many believed that these functions belonged under the control of a single entity, the electric company. To protect the public interest, therefore, the government assumed the responsibility of providing oversight of the industry.

Federal involvement in the regulation of the interstate electricity business began with a law called the Federal Power Act of 1935. Its stated goal was to assure "just and reasonable" power prices. State regulatory agencies were also established that governed intrastate aspects of the business. This system worked well for a long time. Consumers received reliable high-quality electric power at reasonable rates.

But some aspects of the system were problematic. In particular, this regulatory model enabled many utilities to build some very large and expensive projects in ways that were less than efficient. The costs that the plant owners incurred could, with the permission of the appropriate regulatory agencies, simply be entered into the rate structure—in particular, unnecessary costs were often just passed to the customers. Regulating such practices out of existence proved very difficult, to a degree because ineffective regulation was sometimes part of the problem. During the 1970s and 1980s some of the least-efficiently built generating facilities in the United States were nuclear reactors. Some projects ran billions of dollars over budget, and these costs were simply tacked onto the bills of local ratepayers. The ratepayers, of course, had no alternative but to pay.

In addition, the old regulatory structures often failed to provide traditional utilities with sufficient incentives for research and development and for maintaining adequate generating capacity to meet the ever-growing demand for more electricity. With new technology and new ideas, however, a new model for the electric power industry evolved.

In 1996, in response to new federal legislation, the Federal Energy Regulatory Commission issued Orders 888 and 889 with the purpose of restructuring the electric power industry. The new strategy embodied in these orders sought to assure that power prices would be just and reasonable, just as the Federal Power Act of 1935 had, but the method by which it sought to accomplish this goal was radically different from the previous one. The new strategy can be summarized in the following brief excerpt:

> ". . . [Order 888 requires] all public utilities that own, control or operate facilities used for transmitting electric energy in interstate commerce to have on file open access non-discriminatory transmission tariffs that contain minimum terms and conditions of non-discriminating services. . . ."

This is often referred to as the "unbundling" of electric supply services. The effect of the rule together with other changes to federal and state legislation and regulatory policy has been to separate the three functions of electricity production, transmission, and distribution. Unbundling has been accomplished in somewhat different ways and to different degrees in different areas of the country. It remains a work in progress.

The unbundling of the functions of power production and transmission is particularly crucial. The regulation states that any company or consortium of companies that owns an electricity transmission network, by which is meant a system of high-voltage transmission lines, must grant all power producers access to the network in a nondiscriminatory fashion. Conceptually, in the new regulatory environment, the transmission system is often modeled as a highway to which all electricity producers have equal access. Today, provided the power producer can find a buyer and is willing to pay the necessary fees (tariffs), it can use the transmission system to deliver its product to market. In the years since Orders 888 and 889 were issued, there has been tremendous growth in the number of independent power producers. In theory, the competition among power producers to provide the lowest cost electricity will create the necessary incentives to eliminate inefficient practices. Competition among producers should also, in theory, lead to more investment in research and the development of new technologies.

These new competitive electricity markets are often described as deregulated, but this is not entirely accurate. Certainly, the transmission system is still tightly regulated. It is still described as a natural monopoly. Along each transmission corridor there is only one transmission system. It is, moreover, unreasonable to create multiple high-voltage transmission systems along the same transmission corridor. Consequently, in order that the many new generating companies compete on equal terms, the concept of open access to the transmission system must be strictly enforced. This is done by an entity that is often called an independent system operator or ISO. The exact duties of the ISO vary from region to region within the United States, but one responsibility shared by all ISOs involves ensuring that all producers in the regional market have an equal opportunity to sell electricity across the regional grid. One aspect of exercising this responsibility involves routing power through the high-voltage lines in a way that is efficient. An ISO may

ISO New England

To see how the restructured utility industry works in practice, consider ISO New England, a not-for-profit corporation established by the Federal Energy Regulatory Commission to act as the independent system operator for the New England regional power market.

It is important to emphasize at the outset that ISO New England is an independent organization. It owns no power plants and no part of the distribution system, and individual employees are barred from having any financial interest in any participant in the New England power market.

Competitive markets began operation in 1999. Since that time, five New England states—as of 2008 Vermont remains the sole exception—passed legislation requiring utilities to sell their power-generation facilities and purchase electricity on the open market. Within five years after the New England markets began operation, investors had added 9,000 MW of new generation capacity, virtually all of which was in the form of natural gas–fired power plants. Competition had unleashed a building boom but not a burst of creativity and innovation.

Another responsibility of ISO New England is to balance supply with demand. To this end, ISO New England evaluates the state of the system every four seconds. As demand increases, more plants are brought online—lowest-cost producers first. If necessary, higher-cost producers are brought online later. As demand falls, the higher-cost producers are turned off first, and the lowest-cost producers are turned off last. From a technical viewpoint, matching supply with demand in real time is very challenging. Demand and prices fluctuate continually. Not surprisingly, many of ISO New England's 400 employees are computer professionals.

To deliver the electricity to the consumer, ISO New England manages the 8,000-mile (13,000-km) high-voltage transmission system that connects the six New England states. It controls how power is routed through the system in order to maximize efficiency and minimize the possibility of a system overload. To accomplish this task, ISO New England also has the responsibility of evaluating the high-voltage transmission

(continues)

(continued)

system for upgrades. It works with the owners of the system to schedule maintenance and improvements. Anytime a section of the high-voltage system is taken off-line, the way that power is routed throughout the system must be reassessed.

Finally, ISO New England oversees New England's $11 billion-per-year electricity *spot market,* which consists of the day-to-day power transactions upon which utilities depend to meet demand fluctuations that are not met through long-term contracts.

Has this experiment been a success? The system is reliable, but it is not clear that it is cheaper. There is more generating capacity since the markets were restructured, but it is largely in the form of natural gas plants. Natural gas plants were attractive to investors because they are reasonably clean-burning and relatively inexpensive to build. As of 2008 about 40 percent of New England's capacity runs on natural gas. Still more natural gas plants are under construction. But supplies of natural gas have proven to be expensive and prone to disruption. (In 2005, for example, Hurricane Katrina disrupted 16 percent of the nation's natural gas production when it made landfall, damaging much of the energy infrastructure in the Gulf of Mexico and leading to months of sharply higher prices.) An additional 25 percent of New England capacity uses oil, the price of which is very volatile. Much of the rest of the region's capacity is supplied by an aging complement of nuclear reactors. This is what market forces have produced. It is probably too early to declare the success of the restructured electricity markets.

also be responsible for ensuring that there is sufficient generating capacity in the event that there is a surge in demand or a particular generating unit fails to provide the promised electricity. ISOs may also facilitate energy trading, meaning that they may facilitate real-time transactions between electricity buyers and sellers. With respect to ensuring equal access, however, the government, through the Federal Energy Regulatory Commission, is the final arbiter of whether the condition of nondiscriminatory transmission charges, or tariffs, has been satisfied.

Restructuring has fostered competition among many power producers. Each producer has its own suite of technologies upon which it relies to produce electricity. But to the consumer all electricity from the grid is the same; only the price of the product varies. Those who lobbied for the restructuring of the electricity business hoped that increased competition among power producers would encourage innovation in the methods by which electricity is produced, and the consumer would, as a consequence, benefit through lower electricity prices. Legislators and regulators believed that "the free market" would lead to lower prices and higher efficiencies. Whether this has actually occurred is a matter of debate, but these policies have had interesting and important consequences for nuclear power producers.

THE NATURE OF SUPPLY AND DEMAND

The fundamental fact that distinguishes electricity from other commodities is that it cannot be stored. Electricity must be produced exactly when it is needed. In the electricity business demand and production occur simultaneously.

The demand for electricity varies continuously. It varies in different ways on different time scales: hour-to-hour, day-to-day, month-to-month, and year-to-year. Some of the fluctuations in demand that exist on each timescale are predictable and system operators can plan for them, but some are random. Random fluctuations must also be taken into account. The goal is to be certain that there exists enough production capacity to always meet the continuously fluctuating levels of demand.

Some demand patterns are relatively simple to predict. When averaged over several years, for example, electricity consumption

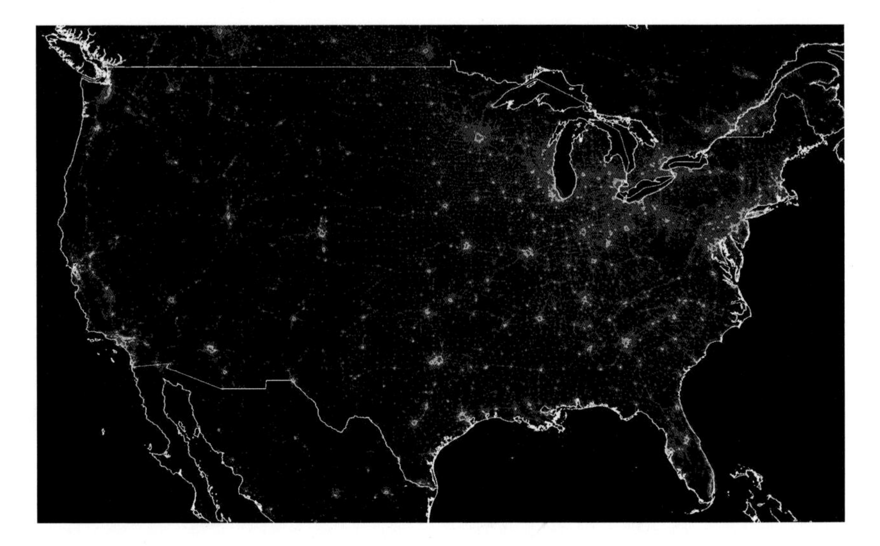

The contiguous 48 states at night. The picture shows where the demand for electricity is greatest. *(NASA)*

has always increased. Over the long-term, electricity usage increases not only in absolute terms—that is, the total amount of electricity consumed increases—it also increases in terms of its relative importance as an energy source. For example, in the United States, in 1970, electricity constituted 8 percent of total energy consumption, but 30 years later, in the year 2000, electricity accounted for 16 percent of the nation's total energy consumption.

Another scale of variation that is fairly predictable is seasonal. During the summer, for example, electricity demand rises, because air conditioning is a very energy-intensive technology. Of course, no one knows exactly which days will be hot—the precise days on which a heat spell will occur is an example of random variation—but everyone knows that the warm days of summer are coming and with them increased demand for electricity. Utilities plan accordingly. There is also predictable day-to-day variation—for example, there is less demand for power on Sundays and holidays, when many businesses are closed, than on weekdays.

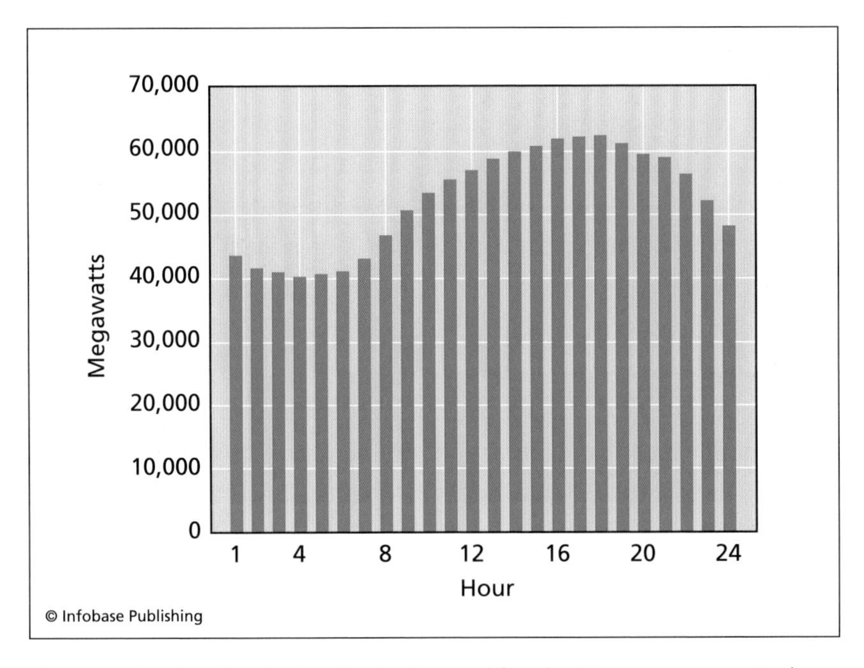

Diagram showing the fluctuation in demand for electric power over a 24-hour period. The base load power level is approximated by the height of the shortest bar. *(Source: North American Electric Reliability Council)*

Finally, there is hour-to-hour variation. Some essential services and certain types of manufacturing operations require power all of the time. Each service area, therefore, has its own characteristic base load, or minimum power-production level, that must be met. Late at night and very early in the morning base load power is often all the power that must be produced to meet the total demand. As the day progresses, however, the demand for power begins to increase. This is called peak power, and utilities must also have strategies in place to produce or buy the additional power needed to meet these hour-to-hour fluctuations in demand. (Sometimes peak power is further divided into intermediate and peaking demand, but here, as in many other texts, all demand above base load will be called peak power.)

Market conditions and the technology by which power is supplied to meet demand depend somewhat upon the quirks of the regional power market that one considers. Following restructuring, for example, some states required that traditional utilities sell all of their generation facilities and purchase all of their power in the marketplace. Some of this power—particularly base load power, the demand for which is highly predictable—is purchased through long-term contracts. (Most base load power is provided by coal, nuclear, and hydropower facilities. Hydropower is more flexible, but coal and nuclear plants generally function most efficiently when allowed to produce power continuously at relatively steady rates for long periods of time.) Peak power can be purchased over the short-term. As demand rises there are, in theory, producers waiting to supply power. The utility will purchase power at the lowest price available, and if there is additional unmet demand after this purchase, the utility will purchase the necessary additional power from the next higher priced producer. The more demand there is, the more power must be purchased and, as a general rule, the more each additional unit of power costs. Because electricity is essential, the utility must continue to climb this "cost ladder" until sufficient supplies have been procured or until all supplies have been exhausted. This market-based model has had some unintended effects.

Because generating companies cannot simply incorporate their costs into the rate structure in the way that traditional utilities routinely did, they are more averse to projects with high *capital costs.* (Capital costs consist of the money that one must spend on a project before it begins to generate revenue.) The cheapest type of large-scale commercial power plant one can build today uses natural gas as a fuel. Natural gas, which was inexpensive at one time, is no longer—the price almost tripled between 1997 and 2005—but the cost of *building* a natural gas–fired plant remains a comparative bargain. Natural gas plants do not require the extremely sophisticated control and safety systems that characterize nuclear plants, nor do they require the expensive antipollution equipment of coal plants. These are some of the reasons that much of the additional capacity that has come on the market in recent years has been in the form of natural gas generating facilities. In fact, in the United States as a whole, from the beginning 1999 through the end of 2002—a period of just four years—investors added 144 gigawatts (144,000 MW) of electrical-generating capacity, and 138 gigawatts of this total, or 96 percent, was in the form of natural gas plants. More recently, the trend toward natural gas has slowed, but it has not stopped.

As more natural gas plants have been constructed, demand for natural gas has risen and so has the price. Many consumers who use the fuel for heating their homes, their hot water, or for cooking have experienced large increases in natural gas prices. Saving consumers money was one of the stated goals of restructuring. It is not clear that on balance this goal has been met. It is easy to point to states with restructured electricity markets, for example, Maryland and Indiana, where electricity costs soared simultaneously with increases in residential natural gas prices. More telling, however, is that consumers in the 17 restructured markets paid on average 30 percent more than consumers in those markets that remained more tightly regulated.

Concerns about the natural gas power plant building boom were expressed November 2, 2005, during testimony by Steven E. Ewing, President and Chief Operating Officer of DTE Energy Gas, who was testifying on behalf of the American Gas Association (AGA) before the U.S. Congress's House Subcommittee on Energy and Air Quality. (The AGA is a major industry group.) The Committee was taking testimony about natural gas and heating oil for American homes. Of special interest to Committee members was the volatile and very high cost of heating oil and natural gas for the residential market. In his testimony Mr. Ewing urged Congress to promote fuel diversity among electricity producers. The preference for natural gas–fired plants by power producers had sharply increased demand for natural gas at a time when supplies remained constant. As a result, prices had increased. He urged Congress to "provide incentives" for producers to turn to coal, nuclear, and alternative fuels and simultaneously lower any regulatory barriers that might impede this transition.

The situation is further clouded by the issue of costs. Restructured markets are supposed to send "price signals" to power producers—by which is meant they are supposed to lose money for bad decisions and earn money for good ones—but there are hidden costs associated with power production that are not addressed by the current system. The market was created so that profits flow toward the lowest cost producer, but little consideration was given to assessing the true costs of producing electricity. The costs of depending on "inexpensive" fossil fuels are, for example, distributed differently than are the costs of nuclear power. With respect to nuclear power, all fuel costs, including the costs of disposing of all of the spent fuel, are an explicit part of the rate structure. By way of example, the cost of the proposed facility at Yucca Mountain will, as a matter of law, be borne entirely by ratepayers through a fee on electricity generated via nuclear power.

The situation is completely different for fossil fuel plants. In this case, the profits associated with operating a fossil fuel generating station are reaped solely by the owners, but because the resulting carbon dioxide is vented directly to the atmosphere, the environmental costs of operating the plant, which appear in the form of higher global temperatures, are distributed among all the inhabitants of the planet. What *are* the costs of power production? And what criteria should be used to identify low-cost producers? These are important questions, but as yet there are no generally agreed-upon answers.

The more recent emphasis on short-term profits in the U.S. markets, a result of the restructuring of the electricity business, has made it unlikely that projects with high capital costs will find investors when other "cheaper" methods of producing electricity exist. The large commercial nuclear power plants currently in operation were probably only possible under the old system of regulated local monopolies. Although the old system required that the owners invest billions of dollars up front—that is, prior to the time that the plants became operational—profits were essentially guaranteed because these capital costs were simply passed to the rate-payers. One of the most extreme examples of an uneconomical construction schedule is the Watts Bar-1 plant in Tennessee, the most recent nuclear reactor to begin operation in the United States. Watts Bar-1 went online in 1996, 24 years after the start of construction. In a restructured electricity market, these types of construction schedules are a thing of the past. But companies have adapted. General Electric's new Advanced Boiling Water Reactor, for example, requires about 39 months to build, and some of these plants have already been built on schedule in Japan.

There is general recognition that the old system under which nuclear plants were licensed was characterized by inefficient regulatory practices on the part of the Nuclear Regulatory Commission. The old procedures—that is, the set of rules that were in effect at the time that the present generation of nuclear plants was being built—allowed interested parties to challenge the design, associated impacts, and construction of each individual plant at virtually any stage of its construction. This was, in part, unavoidable because the technology was new, and all parties were learning by doing. But these challenges, even when unsuccessful, sometimes caused construction delays that were very costly for plant owners. Years ago, opponents to nuclear power could punish a utility—even force a utility to the brink of bankruptcy—not by winning lawsuits, but simply by filing them, thereby delaying the construction and licensing process.

Probably the most spectacular example of this occurred in Shoreham, New York, where local and state opposition to the construction of a nuclear reactor continued throughout the construction period and into the time that the Shoreham facility was granted a license for low-power tests in 1985. The reactor was complete and ready to be put into service when the utility finally succumbed to the pressure and agreed to dismantle the reactor without ever using it to generate useful amounts of electrical power. Billions of dollars were wasted despite the fact that there was no reason to believe that the reactor would not perform as designed. The Shoreham experience demonstrated how financially risky the construction of nuclear power plants had become.

New licensing procedures are in effect today that seek to rationalize the construction of nuclear plants by resolving many of the conflicts about such plants once and for all prior to the onset of construction. The process is further streamlined by the concept of "standard designs." Early U.S. reactor designs were one-of-a-kind. Consequently, each design needed to be evaluated separately. Now the NRC evaluates standard designs—some have already been approved—so that a prospective power producer can choose a reactor design that has already been through the very thorough and lengthy NRC licensing process.

Standardization is a practice that France adopted long ago, in part, because it reduces design costs and makes plant construction more efficient. But standardization also improves safety because operating experience with one reactor is directly applicable to all others of the same design.

NUCLEAR POWER IN THE UNITED STATES TODAY

As of this writing there are 104 commercial nuclear power stations operating in the United States, the largest number of nuclear power stations in any nation in the world. The 104th, the Brown's Ferry 1 nuclear plant, which ceased operation in 1985, returned to service in June of 2007. (There are numerous other reactors used for scientific research or by the Department of Defense. These are not considered here.) The commercial reactors, all of which are light

Nuclear Regulatory Commission headquarters, Bethesda, Maryland

water reactors, provide about 20 percent of the electricity produced in the United States. They currently operate, on average, in excess of 90 percent of their capacity—that is, they operate at full power more than 90 percent of the time—a remarkable improvement over the 56 percent capacity factor that marked the industry average in 1980. In addition to increases in capacity factors, many plants have already obtained or are in the process of obtaining permission to operate at higher power levels. Increased capacity factors and the higher operating power levels explain why, despite the fact that not a single new plant has been ordered since the 1970s, the amount of electricity generated each year by nuclear power has slowly increased.

Increases in the cost of oil and gas together with increases in efficiency at the nuclear plants themselves have made nuclear, once one of the more expensive generating technologies, a comparative bargain in two ways. First, the average price of electricity generated using oil and natural gas now tends to be higher than that generated by nuclear power. Second, the price and supply of oil and gas have proven to be unpredictable. Affected by hurricanes, ever-growing demand, and political crises, gas and oil prices have become increasingly volatile. Nuclear fuel is not just comparatively inexpensive, but electricity produced via nuclear power is also stable in price. Nuclear power enables producers to confidently sign long-term contracts that protect consumers, industrial and residential alike, from large and unpredictable price fluctuations. (The volatile nature of oil and gas supplies is reflected in the fact that oil and natural gas suppliers have become increasingly reluctant to sign the long-term contracts that assure consumers of price stability.)

Safer, simpler, less expensive nuclear power plant designs relying upon more sophisticated safety features are now available. Some of these designs—the Advanced Boiling Water Reactor design already deployed in Japan, for example—have been licensed for use in the United States. The Nuclear Regulatory Commission has worked hard to rationalize its procedures for issuing licenses, a change that may help to reduce investor uncertainty. A growing awareness that the current rate of consumption of fossil fuels—and that rate is accelerating—is causing widespread environmental changes that will have dire consequences for many is still another factor that makes nuclear power an increasingly attractive alternative. In fact, a tax on greenhouse gas emissions may make nuclear energy (the only large-scale generating technology that produces zero emissions) the technology of choice for many power producers.

Working against additional investments in nuclear power is the concern that the technologies used to enrich and reprocess fuel can be harnessed to produce nuclear weapons. And there is the failure of the federal government thus far to open a waste repository for spent fuel or to adopt reprocessing technology. There is also a perception among many that nuclear power is inherently dangerous and that there exists no safe solution to the problem of disposing of the wastes generated by these plants.

During the early years of the restructured markets, the high capital costs of building reactors was a deterrent to constructing new plants, but there is ample evidence that this situation is changing. For the reasons already given, new nuclear plants are becoming increasingly attractive investment opportunities when compared with fossil fuel plants, the only other technology currently capable of meeting the enormous and continually increasing demand for electricity.

In a speech on February 16, 2006, Clay Sell, Under Secretary of Energy, asserted that global energy demand will increase by 50 percent during the next 25 years and expressed the belief that demand will double in 50 years. Much of this new demand, he said, will be met with nuclear power. He noted that Asian nations, where much of the growth is occurring, are beginning to invest heavily in nuclear power. It is the stated policy of the United States, he said, to influence the development of this technology, to ensure that it is as proliferation-resistant as possible, and to create designs that are as safe as possible. To do that, the United States must once again become the leader in commercial nuclear power. That would require a vibrant nuclear industry within its own borders. According to Mr. Sell, the government's goal is to have 300 nuclear reactors in service by the year 2050.

Nuclear Energy and National Policy

Throughout the 1980s and 1990s nuclear energy was widely perceived as inherently dangerous and dirty, and it was difficult to find anyone outside of the nuclear power industry willing to publicly support the technology. This has changed. Today the situation is more nuanced. Many governments and self-described environmentalists are again looking at nuclear energy as a way of safely generating large amounts of emissions-free electricity. Faced with the tremendous economic growth occurring in Asia together with the steadily growing economies of many developed countries, many of those interested in finding the best way to meet this enormous demand for electrical energy now recognize that, while conservation efforts are valuable, there is no conceivable way of conserving sufficient energy to eliminate the need for large increases in electricity production.

It is not possible, given the many different viewpoints on the value of nuclear power, to comprehensively describe all arguments for and against nuclear power. Instead, this chapter begins by briefly summarizing some of the major ideas for and against nuclear power and proceeds to describe how these ideas have affected the energy policies of three nations: Germany, the United States, and France.

SOME ARGUMENTS FOR NUCLEAR POWER

Argument #1: Nuclear Energy Can Supply Sufficient Power to Meet Demand

A typical nuclear power plant is, relative to most other types of power plants, a large producer of electrical energy. The United States, by far the world's largest economy, currently meets approximately 20 percent of its electricity needs with only 104 nuclear power generating units. Not only do these power stations produce a great deal of energy when they are in operation, they are in operation most of the time. The capacity factors of nuclear power plants—that is, the percentage of time at which each operates at full power—average approximately 90 percent, higher than most other power-generation technologies. Finally, nuclear energy enables designers to reliably match electricity supply with electricity demand, because it is not enough that a generating unit produce power, it must produce power when there is a demand for it. (Consider the problem of attempting to meet demand with solar or wind power. Simply installing additional solar and wind units does not ensure that there will be sufficient supply when it is needed. The electricity may be needed at times when the Sun is not shining and the wind is not blowing. The Sun, after all, does not shine at least half of the time, and the wind blows only intermittently. Solar and wind units produce power—or not—independently of demand.)

Argument #2: Nuclear Power Provides Emissions-Free Electricity

Nuclear power plants do not depend upon combustion to produce electricity, so they produce none of the products of combustion.

Carbon dioxide, one of the principal by-products of the combustion process, is an important causative agent of global climate change. To see the scale of the numbers involved, imagine replacing a single one-GWe nuclear plant with one or more coal-burning power plants. To produce the same amount of electricity as the nuclear power plant, one must burn roughly 2.5 million tons of coal each year, a very large environmental burden.

Argument #3: The Electricity Produced by Nuclear Power is Economically Competitive

This argument depends on two observations: First, in the United States all the costs of nuclear energy, from the costs of mining the uranium to the costs of disposing of the spent fuel, are borne by the consumer of nuclear energy. Coal and natural gas plants, the other two major sources of electricity in the United States, also produce waste, but the consumer is not charged the costs of disposing of the CO_2 produced by these plants, and carbon dioxide is one of the principal waste products of combustion. The cost structure of nuclear energy is, therefore, different from the cost structure of fossil fuel technology. Even with this provision, electricity produced from nuclear energy is cheaper than that produced by burning natural gas or oil, and it is competitive with that of coal. The second observation on which this argument rests is that the cost of power produced by nuclear energy is highly stable. Because nuclear fuel releases so much energy in the fission process, fuel cost is only a small part of the cost of producing power with nuclear energy. Since the 1970s, the price of uranium has fluctuated widely, but this has had only a minor effect on the price of the electricity produced by nuclear power because unlike the cost of natural gas, for example, nuclear fuel costs comprise a small fraction of the cost of electricity production. Price stability enables businesses and homeowners alike to successfully plan their future power expenditures.

Argument #4: The Problem of What to Do with Spent Nuclear Fuel Has Been Solved

This cannot be strictly true—at least not in the United States, where the fate of the geological disposal site at Yucca Mountain is uncertain—but there are many who believe that a geologic repository is a practical way to isolate spent fuel. Sweden, Finland, France, and the United States, for example, all have plans to use geologic repositories. Moreover, new technologies can reduce the amount of waste that must be placed in the repositories and can even reduce the length of time that the waste will remain dangerous. The plutonium and uranium present in spent fuel can now more safely be reprocessed and used again, and other heavy metals that remain radioactive for millennia can be destroyed in specially built reactors so that only waste that is highly radioactive for centuries rather than millennia will remain to be placed in a repository, a much more tractable problem.

Argument #5: New Designs Solve Most of the Problems Associated with Nuclear Power

No matter what one thinks of the reactors in use today in the United States, no additional plants will be constructed using these designs, because the science and engineering associated with nuclear power plant design, construction, and operation have dramatically improved since the first U.S. nuclear plant building boom ended decades ago. New designs, some of which are already licensed by the NRC, and new reactors already under construction in Europe and Asia, are safer than their predecessors; they are less expensive to build and operate, and they offer the promise of reasonably priced electricity with far fewer environmental consequences than other currently available technologies. Future designs, the so-called Generation IV plants, offer still more advantages.

SOME ARGUMENTS AGAINST NUCLEAR POWER

Argument #1: Nuclear Power Plants Are Dangerous

Nuclear power depends upon fission, a phenomenon that is inherently unstable. Small changes in the operation of a nuclear reactor can quickly lead to large and dangerous power fluctuations. The complexity of the technology used to control nuclear power plants makes it difficult to respond effectively to these power surges, and the history of nuclear power, especially the Chernobyl accident, is sufficient proof that these plants cannot be operated safely.

Argument #2: Nuclear Power Plants Increase the Likelihood of Nuclear Proliferation

Nuclear weapons use either highly enriched uranium or plutonium-239. First consider the problem of producing slightly enriched uranium for use as fuel: The technologies used to manufacture slightly enriched uranium for use in commercial reactors can also be used to manufacture highly enriched uranium for use in nuclear weapons. Consequently, enrichment technology, which is necessary to manufacture fuel for most current designs of nuclear reactors, poses a substantial proliferation risk. Second, consider the problem of spent fuel. Substantial amounts of plutonium are created within the fuel of an operating reactor. The plutonium can be separated during reprocessing to produce the material needed to make a plutonium bomb. As with the fuel manufacturing stage, the reprocessing stage also enables the user to produce the material necessary to make a nuclear weapon.

Argument #3: The Electricity Produced by Nuclear Power Plants Is Expensive

Nuclear energy has benefited from significant government subsidies both in terms of government-funded research and, for example, policies that limit the liability of nuclear plant owners in case of a power plant disaster. If the costs of these programs were fairly reflected in the electricity rates paid by the consumer for nuclear-generated electricity, electricity bills would be much higher.

Argument #4: The Problem of What to Do with Spent Nuclear Fuel Has Not Been Solved

The plutonium and uranium in spent fuel remains radioactive for very long periods of time—much longer than the recorded history of the human race. One cannot, therefore, be sure that the strategies currently envisioned for placing the spent fuel deep underground in specially built repositories will safely retain the radioactive material for sufficiently long periods of time. This generation has a responsibility to future generations to find a more effective solution to the problem of radioactive waste than to simply bury it deep underground in a place from which it may well escape long before the contents of the repository have decayed into stable isotopes.

Argument #5: Nuclear Power Is Unnecessary

Other "greener" technologies exist that can, in theory, generate similar amounts of power. These technologies are safer and produce less pollution than nuclear power. The reason that these technologies currently generate very small amounts of power in comparison with nuclear power is that they have not benefited from the same amount of government-funded research and other subsidies. Alternatives to nuclear energy include but are not limited to solar energy, geothermal energy, wind energy, tidal energy, and biofuels.

Evidently, there are two sides to the debate, but from a scientific or engineering perspective it would be irrational to assert that all points of view are equally correct. Because some of these assertions involve matters of science and engineering, they are subject to verification. In some cases, therefore, there are ways to separate what is true from what is false. But some of these assertions are less open to unequivocal proof, and in any case, politics is also important as the following three cases make clear.

NUCLEAR POWER POLICY IN GERMANY

West Germany assumed an important leadership role in the development and implementation of commercial nuclear energy beginning in the 1960s. Government policies toward nuclear power have, however, changed, and Germany now hopes to have every reactor in the country shut down by the year 2020.

The policy of the West German government toward the development of nuclear power began with the 1959 Atomic Energy Act, which, among other objectives, aimed to promote the use of nuclear energy. Although West Germany never deployed any pebble bed reactors, a technology that originated in that country, it did build and continues to operate a number of light water reactors. In 1970, for example, it was operating eight light water reactors; it operated 24 light water reactors in 1985; 20 years later that number had decreased to 17 nuclear reactors in operation—11 pressurized water

Nuclear power plant in Stade, Germany *(Boereck)*

reactors and six boiling water reactors. All of these reactors have excellent safety records.

Prior to the unification of the two Germanys, East Germany also deployed a number of nuclear reactors. These were all of Soviet design and included some that were of the same type as those at Chernobyl. While these reactors had been operated largely without incident, it is generally acknowledged that they were not of the same caliber as those in the West. The accident at Chernobyl heightened concerns about their safety. In 1990, a unification treaty was passed by both the East and West German governments, and the newly united Germany found itself in control of 11 Soviet-designed reactors that were either in operation or under construction and 21 reactors of Western design located in the former West Germany. During the early 1990s, work was abandoned on all East German nuclear reactors under construction and all the Soviet-designed reactors in operation were shut down.

During the late 1990s, government policy toward nuclear energy began to change. The ruling Social Democrats under the leadership of Gerhard Schröder together with the German Green Party began a prolonged and difficult series of negotiations with Germany's major energy producers, with the goal of shutting down all of the remaining nuclear reactors in Germany. In 2000, an agreement was reached: No new nuclear plants would be licensed to operate in Germany and each plant in operation would be limited to an operating life of 32 years. In addition to the restrictions on reactor construction and operating life, it was further agreed that Germany would halt all additional shipments for fuel reprocessing by 2005. (Spent fuel was being reprocessed in France.) Nuclear plant operators also agreed to construct interim storage facilities at the sites of the reactors and hold the spent fuel at those sites until a decision is made regarding the final disposal of the spent fuel. These agreements were formalized in the new Atomic Energy Act of 2002.

What was the motivation for this change in policy? Jürgen Yritten, Germany's federal environment minister, described the decision as "the logical response to Chernobyl," and much of the justification of the change in policy has been expressed in terms of public safety and a general concern for the environment. In 2001, Wolfgang Renneberg, director general for nuclear safety in the Federal Ministry for Environment, Nature Conservation, and Nuclear Safety, in a speech in Madrid, Spain, gave three main reasons. First, the government had reassessed the risks posed by nuclear power. Although the safety and reliability records of these plants was very good, the German government decided that the damage that could be done by a worst-case nuclear accident was so severe that, despite the very low probability of such an accident, all the plants should be shut down as soon as practicable. Second, although Germany had spent a great deal of time and money studying the possibility of geological disposal in a formation called the Gorleben salt dome, it remained unconvinced that this was a workable solution. Because spent fuel was perceived as a "burden for future generations of people," it had been decided to stop generating additional spent fuel by shutting down the reactors where it is produced. Third, Germany reassessed the proliferation risks posed by nuclear power. By putting an end to fuel reprocessing and reactor operation, the German government believed that it was reducing the stockpile of material that could be used to build nuclear weapons and consequently improving its national security.

In 2002, at the time of the passage of the new Atomic Energy Act, nuclear energy supplied about one-third of Germany's energy requirements. About 50 percent of its electricity was obtained by burning coal. Natural gas was also an important source of energy. The fuel mix is important because in addition to its decision to phase out nuclear power, Germany has committed itself to the goal of decreasing its greenhouse gas emissions and, in particular, to minimizing its reliance on coal as a primary energy source. Therefore, Germany cannot simply construct more coal-burning plants to compensate for the shortfall in electricity production caused by eliminating nuclear power.

In order to accomplish both goals—to decrease greenhouse gas emission and to eliminate nuclear power—Germany is forced to rely on what are often described as renewables, which are hydropower, wind, solar, geothermal, tidal power, biomass, and possibly some yet-to-be-deployed technologies. Hydropower in Germany as elsewhere is limited by the availability of sites. Essentially all of the good sites in Germany have already been developed, leaving little room for expansion. There is not enough arable land in Germany to replace a significant fraction of the electricity produced by nuclear power with biomass. There are, in fact, significant difficulties with the large-scale implementation of most of the other technologies on the list. Consequently, the German government has invested heavily in wind turbines and photovoltaic solar power. There are tens of thousands of wind turbines scattered throughout Germany, with a preponderance of these located in the north, where the wind supply is best. German wind turbines enjoy heavy subsidies in the form of a law that requires utility companies to purchase power from renewable energy sources even when such power is more expensive than that produced by more conventional sources. As the number of wind turbines has increased and the total price of the subsidy has risen, support for the use of wind technology has weakened. Photovoltaic power is also developing rapidly, but weather patterns in Germany as well as the country's northerly location place severe constraints on the total amount of power that can be generated with this technology. It is also a very expensive technology to use. It is not yet clear how much of a contribution photovoltaics can make to the mix of generation technologies on which Germany relies. Nor is it clear how much large-scale power generated from the Sun would cost. There are some who claim that with present technology Germany cannot possibly decrease its commitment to nuclear power and at the same time cut greenhouse gas emissions on the timetable to which the government has already committed itself. These objections were dismissed by the Schröder government as old-fashioned thinking, but the situation is not so clear. Germany is now embarked on a coal-fired power plant building boom.

Time will tell whether the German government sees more deeply into the problem of electricity production than its critics or whether it is simply trying to legislate physics. As of 2008, officials in the Christian Democratic Union government of Angela Merkel are reevaluating their decision to shut down the nation's reactors on schedule in the face of doubts over the reliability of Russian natural gas imports.

NUCLEAR POWER POLICY IN THE UNITED STATES

To appreciate the current state of nuclear power in the United States, it is important to understand that decision-making authority is broadly distributed among different branches of government. Most, but not all, regulatory authority resides with the Nuclear Regulatory Commission (NRC). There are some areas, for example, where the Environmental Protection Agency (EPA) has responsibilities that overlap those of the NRC. But as its name implies the NRC is a regulatory body; policy decisions are made elsewhere. Individuals and citizens seeking to influence policy may contact their elected officials on the federal, state, or even local levels, or they may attempt to make their views known through the courts, and they often do. Some observers assert that this way of distributing authority ensures that a policy will be approved only if it enjoys broad support. Others describe the situation by saying that the current situation only guarantees that everyone has the right to say no. As a consequence, different governmental entities often work at cross-purposes. The German experience, in which the government formulated a single nationwide policy requiring the early closure of all

Diablo Canyon power plant, San Luis Obispo County, California *(Marya)*

reactors in the nation and an end to the deployment of new reactors, probably could not happen in the United States, where authority is considerably more diffuse and decisions, when they are made at all, are subject to frequent reversal.

With respect to the regulation of commercial nuclear reactors, the Nuclear Regulatory Commission (NRC), established in 1974 by President Gerald Ford, has very broad responsibilities. Its duties include regulatory oversight of all commercial nuclear reactors. There are three areas that receive particular attention: (1) reactor safety, which involves avoiding accidents and mitigating the consequences of any accidents that might occur; (2) radiation safety, which involves establishing radiation exposure levels for plant workers and the general public and ensuring that reactor operators adhere to these standards; and (3) safeguards, which involves protecting each plant from sabotage and other threats to its physical integrity.

The NRC also regulates new reactor siting, construction, and operation. To accomplish these goals, the NRC requires applicants to obtain approvals in the form of certificates and licenses. If the applicant satisfies the NRC that the relevant criteria have been met, then the NRC will issue the required approvals. In an indirect acknowledgement that past regulatory practices were less than ideal, the NRC now asserts that it is working to "make new licensing reviews more effective and efficient and to reduce unnecessary regulatory burden on future applicants."

Regulatory questions are often quite technical. Obtaining a standard design certification for a new reactor design, for example, takes years, and the problems associated with evaluating reactor designs have become more challenging for NRC staff as a wider variety of designs are submitted. Evaluating claims by manufacturers requires ever-increasing levels of expertise on the part of NRC staff. The situation is further complicated as the NRC seeks to satisfy multiple and sometimes conflicting regulatory requirements. (As noted in chapter 6, the NRC also has responsibility for other aspects of commercial nuclear reactor operation, such as the safe disposal of spent fuel.)

But the NRC regulates in a legal environment determined by Congress, the president, state and federal courts, and state and local governments. Authority is evidently widely distributed. The result is a system that is sometimes incapable of making policy decisions. This is best illustrated by the history of the waste repository at Yucca Mountain.

In 1984, the federal government announced its intention to spend $500 million at each of three Western locations: at Yucca

Mountain, at a site in Deaf Smith county, Texas, about 30 miles (48.3 km) west of Amarillo, and at the Hanford Nuclear Reservation near Richland, Washington, to determine the most suitable site for a nuclear-waste repository. The goal was to begin accepting waste by 1998. (There were also secondary sites that were to be considered, one of which was in Louisiana.) The law that initiated the search for the best site also contained a novel provision that enabled the governor of the state in which the proposed facility was to be located to veto the proposal. The veto could only be overridden by both houses of Congress. Without waiting for any studies to begin, the governors of Nevada and Texas both announced their opposition to the plan to initiate studies of the locations in their respective states, and the governor-elect of Washington expressed reservations about the possible choice of the Hanford site.

Often forgotten today is that the enabling legislation for the western sites also contained a plan to build a second nuclear repository in the eastern United States. Seventeen eastern states contained geologically promising sites, primarily granite and shale deposits, which offered the possibility of isolating the highly radioactive spent fuel indefinitely. (Keep in mind that most of the high-level waste destined for the western repository was generated in the East, and that many scientists identified large granite deposits as possibly the best place for a nuclear repository.) In 1985, in response to determined opposition to the western sites, the Department of Energy announced a plan to move forward the selection process in the East. Governor Madeline Kunin of Vermont, a state with large granite deposits and a commercial nuclear reactor, called upon the governors of all the other possible Eastern sites to urge the Department of Energy to drop plans for an eastern repository and instead depend upon aboveground storage. In May 1986, Department of Energy Secretary John S. Herrington announced that no consideration would be given to eastern sites until the mid-1990s. He made the announcement ostensibly because of good progress on identification of the western site. In 1986, Nevada filed five lawsuits against the Yucca Mountain project.

In 1987, conferees of the House and Senate agreed to leave the studies of the best site for a nuclear-waste repository incomplete, and simply designate Yucca Mountain as the sole candidate for a nuclear-waste repository. Senator Bennett Johnston (D–La.), announced, "It's fair to say we've solved the nuclear-waste problem with this legislation." The Yucca Mountain decision received considerable support because it eliminated even the possibility of an eastern repository and removed all other western sites from consideration, as well as Senator Johnston's home state. The provision became law. "The problem with nuclear waste," asserted Senator Johnston, "has never been scientific. It's always been emotional and political."

In January 1989, the Department of Energy (DOE), faulted by the NRC for poor management and shoddy science, postponed the date that Yucca Mountain could begin accepting wastes to 2003 at the earliest. In November 1989, in response to further criticism by the NRC, the EPA, the U.S. Geological Survey, and outside experts, the DOE announced that it would begin anew the study of the suitability of Yucca Mountain. In December of that year the state of Nevada denied the DOE permission to drill at Yucca Mountain and filed yet another lawsuit to prevent construction at the site.

Meanwhile owners of nuclear power plants became increasingly restive about the large stockpiles of spent fuel accumulating at their reactor sites. In 1992, utilities and state officials in Florida and Minnesota accused the DOE of squandering ratepayer money in an increasingly ineffective effort at Yucca Mountain. One year later, DOE secretary Hazel O'Leary announced that the earliest date that Yucca Mountain could open would be 2013. Nevada continued to oppose Yucca Mountain during these years. An important argument was that the geology of Yucca Mountain is incapable of containing the radioactive wastes for the required time period.

Lawsuits and studies continued unabated. In June 2001, Senator Tom Daschle (D–S.Dak.) stated that as long as the Democrats were in charge of the Senate, Yucca Mountain would not open. Eventually, in response to assertions that Yucca Mountain geology alone cannot contain the radioactive material it was being designed to receive, the DOE policy shifted from one that emphasized geologic features to one that placed increased emphasis on the value of engineered barriers—barriers that it estimated would contain the waste for at least 11,000 years. This shift in DOE emphasis caused the state of Nevada to argue that if the engineered barriers are that efficient then there was no need to use a geologic repository at all. In December 2001, Nevada filed yet another suit against the federal government *because* the DOE had changed emphasis from geologic barriers to engineered ones.

Meanwhile, the tenor of the discussion—already more than 20 years old at that point—changed with the attacks of September 11, 2001. Security analysts recognized that although the reactors are relatively safe from attack—they are enclosed in sturdy containment structures—the pools holding the spent fuel are more vulnerable and contain much more radioactive material than the reactor itself. Spencer Abraham, secretary of the DOE, introduced a new rationale for the Yucca Mountain site when he said, "We should consolidate the nuclear wastes to enhance protection against terrorist attacks by moving them to one underground location that is far from population centers," an indirect reference to Yucca Mountain. On February 2, 2002, four years after it was to have opened, the DOE finally recommended to the president that a repository be built at Yucca Mountain. On April 8, 2002, Nevada governor Kenny Guinn vetoed construction at Yucca Mountain. He was overridden by the House two weeks later and by the Senate in July of 2002. The president approved the site that same month.

One might think that at this point the controversy would be over, but in July 2004, the U.S. Court of Appeals for the District of

Columbia ruled in response to a lawsuit by the state of Nevada and a group of self-described environmental organizations that 10,000 years was too short a time for which to plan since the DOE's own analysis suggested that maximum emissions of radioactivity from the site would occur only hundreds of thousands of years after the site was closed. In response, the EPA issued revised health standards for the first million years of use of the repository. Nevada objected. In February 2006, DOE Secretary Samuel W. Bodman admitted that he no longer had an estimate for when Yucca Mountain could begin to accept radioactive waste. The DOE had by this time been served with court orders to begin paying penalties to nuclear operators for failing to accept nuclear wastes on schedule, and at the same meeting at which Bodman appeared, Nils J. Diaz, chair of the NRC, predicted that the nation would shift to aboveground interim storage and fuel reprocessing. The history of Yucca Mountain is representative of how U.S. nuclear power policy evolves.

NUCLEAR POWER POLICY IN FRANCE

French interest in nuclear power began in 1945 with the founding of the CEA (Commissariat à l'énergie atomique), the atomic energy commission of France. Shortly after the war, France began experimenting with a low power heavy water reactor. In 1956, France began development of the first French-designed nuclear reactor, and in 1964 its first commercial nuclear reactor was placed in operation.

In the late 1950s, commercial nuclear power in France received much less attention than the development of a French nuclear arsenal—the first French nuclear weapon, a plutonium bomb, was detonated in 1960—but research and development priorities shifted in 1973, following major oil price increases instituted by the Organization of Petroleum Exporting Countries (OPEC). Prior to the 35th OPEC conference in Vienna, Austria, in September and October of 1973, crude oil was trading at about three dollars per barrel. This is, of course, very cheap by contemporary standards, but it

Nuclear power plant, Catternom, France *(Stefan Kühn)*

was also cheap by the standards of the time—so cheap that oil was routinely burned to produce electricity. Because oil generally burns more cleanly than coal, some utilities even converted coal-burning facilities to oil. This began to change after the Vienna conference, when OPEC members raised prices by 70 percent. Shortly thereafter, in December 1973, OPEC members again raised prices, this time by 130 percent, and, in order to punish the United States and the Netherlands for their support for Israel in its 1973 war with its neighbors, some members of OPEC placed a short-term embargo on these two nations. The result was a severe shock to the world economy. Over the next seven years the price of crude oil increased tenfold as a result of price manipulation, political instability, and war among OPEC members.

In response to these new economic challenges, France switched from fossil fuel technology to nuclear power as a way of generating electricity. In 1970, France had only eight small nuclear power plants generating 1,696 MWe. By 1990, it was operating 56 nuclear power stations generating a total of 55,998 MWe. In addition to the rapid expansion of its nuclear capacity, France sought to standardize reactor design and operation, with the result that France's complement of nuclear reactors is now extremely homogeneous. In 2007, France operated 59 nuclear power plants consisting of 58 PWRs, operated by the national utility, Electricité de France, and one fast breeder reactor, which Electricité de France operates in conjunction with the CEA.

France has come to depend upon nuclear power more than any other major industrialized nation. It is second only to the United States in terms of its total electric output from nuclear power, and as a percentage of its total power output, France has long met in excess of 75 percent of electricity demand through its nuclear power plants, one of the highest percentages in the world. (Although substantial year-to-year variation occurs, much of the remaining non-nuclear contribution comes from hydroelectric power. France has, therefore, learned to produce almost all of its electricity without fossil fuels.) France reprocesses its spent fuel as well as the spent fuel of other nations. It is the major exporter of electricity in the European Union—exporting electricity even to those nations that claim to be against nuclear power—and the safety record of its nuclear industry is excellent.

France's implementation of nuclear power contrasts sharply with that of the United States. The French power generation industry is highly centralized, exactly the type of structure that the United States government has worked hard to dismantle, and French decision-making authority regarding the deployment of its reactors is similarly centralized. As a consequence, when, during the early days of its nuclear power industry, France decided to standardize its complement of reactors, it had the governing and regulatory structures in place to ensure that this goal was met. Standardization is the reason that France uses only pressurized water reactors built by Framatome ANP, a subsidiary of the AREVA consortium. The goal of standardization is to more efficiently construct and operate its reactors. By acquiring expertise in a single design, France is, in theory, better able to monitor, repair, and more safely operate all of its reactors, because operating experience with one reactor is directly applicable to others. France has accumulated over 1,500 years of reactor experience with its reactors and has only recently begun to deploy a new reactor design, the European Pressurized Water Reactor (EPR), a so-called Generation III design, with more sophisticated passive safety features, a higher thermal efficiency rating, and a larger power output.

France has, in common with the United States and several other nations with a nuclear power industry, decided on a geologic repository to store its nuclear wastes. But the French system produces much less waste per reactor than the American because, as mentioned previously, France reprocesses its spent fuel. Consequently, most of the heavy metals are removed from the waste; plutonium and uranium are reused as fuel, and most of what is left, although initially highly toxic, is dangerous only for centuries rather than millennia.

The result of these efforts—already enumerated in chapter 1—is that France has made dramatic progress in reducing greenhouse gas emissions, far more progress than any other major developed nation, even as its country and population have continued to grow.

Conclusion

Attitudes toward nuclear power have changed radically since the first reactors were deployed. Once regarded uncritically as the power source of the future, during the 1980s many came to perceive commercial nuclear power as dangerous, dirty, and not worth the risk. Today attitudes toward nuclear power have become, in many cases, considerably more nuanced as thoughtful people consider the problem of how to reduce greenhouse gas emissions and other environmental hazards while simultaneously providing reliable and reasonably priced electricity to the billions of people who need it.

Meeting the ever-increasing demand for electricity will not be easy. The solutions are often very technical. Producing sufficient electricity at a price people can afford while simultaneously minimizing the associated environmental impacts will require creative thinking from engineers and scientists. But the policies required to implement the technical solutions will be the result of political processes, and to formulate successful policies, politicians will require input from a well-informed citizenry.

Nuclear power can meet this demand now. This much is beyond dispute. Whether nuclear power should be used to meet this demand is a separate question. If one is inclined to answer yes, one should know what this decision would entail. In particular, one should know the limitations of this technology and the problems that accompany its use. If, however, one is inclined to answer no, one should be prepared to answer the question, if not nuclear power, then what? While there are many technologies capable of producing small amounts of electricity, no one has, so far, found a way to harness them to produce large amounts of electricity when and where it is needed. Currently, very few technologies exist for producing large amounts of electricity on demand. With the exception of nuclear and hydroelectric plants, all large-scale options currently entail the burning of vast amounts of fossil fuels.

Producing sufficient electrical power without unduly disrupting the environment will be one of the great problems of the first half of the 21st century. There is currently an almost continuous debate in the media about the best ways to produce more electricity. Nuclear power is coming increasingly to the fore. Many claims about nuclear power are exaggerated; some are demonstrably false; others have real merit. As citizens we have a responsibility to become informed and participate in this debate. Our futures depend upon it.

Afterword: An Interview with Harold Denton: On the U.S. Nuclear Power Industry

Harold Denton is a former director of the Office of Nuclear Reactor Regulation who has been involved with nuclear power since his student days at North Carolina State University, the first university in the country with a research reactor. Upon completion of his college studies, Mr. Denton worked as a reactor physicist at the Savannah River Plant, which was one of the U.S.'s principal sites for the production of nuclear weapons material. In 1963, he joined the Atomic Energy Commission (AEC), the forerunner to today's Nuclear Regulatory Commission and Department of Energy. He came to national prominence and received widespread acclaim for his work under pressure as the chief federal official on the scene at the 1979 accident at Three Mile Island (TMI). In his capacity as director, he has licensed over 40 nuclear power plants, more plants than anyone else in the world. He is one of the principal figures in the history of commercial nuclear power. The following interview took place on February 16, 2007.

Harold Denton during his time at the NRC *(Nuclear Regulatory Commission)*

Q: At one time the Nuclear Regulatory Commission (NRC) had about 200 applications to build nuclear plants. That stream of applications eventually dried up. In the United States no nuclear plants have been ordered for decades. What do you think were the principal contributing factors for the change?

A: When I joined the Commission in 1963, John Kennedy was president, and there were about eight power reactors in operation and four others under construction. Most of these were relatively small demonstration plants that had received government subsidies. (The Commission issues licenses to build, construct, and operate plants based on protection of public health and safety and the environment.) A year later, the General Electric Company announced that it would build the Oyster Creek plant in New Jersey at a fixed price. That plant was heralded by the nuclear power industry as "an economic breakthrough." Applications increased rapidly afterward, totaling 55 reactors in the next three years. We, in the NRC regulatory staff, were swamped with the large number of applications for plants of increasing power levels and complexity. In my opinion, the industry grew too large too fast. At one point, the Commission staff instituted a practice of docketing only one application per week. When I became director of the Office of Nuclear Reactor Regulation in 1978 there were about 230 applications under review.

By 1975, after the energy crisis had slowed the projected demand for electricity considerably, a number of utilities found themselves with more plants being planned or under construction than they needed. For example, at the Shearon Harris plant site near Raleigh, North Carolina, only one of four planned reactors was completed. A number of other utilities also cancelled plants and deferred construction. So it was really the energy crisis that began the slowdown in plant orders. Also, mounting costs and delays were a factor. Then, the TMI accident further convinced some utilities not to finish plants because of the financial uncertainties it posed. At that time, some in Congress were calling for a ban on new plants. There are now 104 plants in operation, the construction of all of which started at least thirty years ago. These plants are operating very safely and reliably today.

Q: What about the Nuclear Regulatory Commission's own licensing procedures? There has been some attempt to change them, but do you think that for awhile these procedures were overly time-consuming and made the siting and licensing of the plants uneconomical?

A: That was a constant industry complaint the whole time I was in the NRC—that regulation was too intrusive and time-consuming. Nuclear power is certainly one of the most heavily regulated industries in this country, comparable to the regulation of airplanes. As large new plants were built, came online, and began to operate, unanticipated safety issues arose. The Commission was constantly adding requirements in technical areas. Also, there was little standardization of designs. By comparison, France selected as their standard model a pressurized water reactor designed by Westinghouse. French President Charles de Gaulle supposedly once remarked that the United States had only one type of cheese but 100 types of reactors while France had 100 types of cheese but only one type of reactor. In the U.S. each utility makes its own decisions about its power sources.

Costs went up greatly after the TMI accident. I estimated at one time that the total cost to the country of the new regulations required after TMI was around $5 billion. In general, these additional costs were not for major hardware such as pipes, pumps, and valves, but for improvements in "human factors" such as the training and qualifications of operators, new and improved operating procedures, improved instrumentation, and much increased attention to the ergonomics of the control room. When TMI occurred, there were only a few simulators in the country. Each of the four reactor vendors had a generic simulator near their headquarters. One of the first things the staff did was to require that all plants have on-site simulators, which would replicate exactly each plant's control room and system behavior.

A simulator at that time cost millions of dollars per plant. But with on-site simulators, plant operators could be rotated in and out of simulator training every six weeks. New operators could be trained on the operation of the plant. As simulation power has increased with increased computing power, simulators have become better and better at reproducing the plant's behavior, providing very realistic training for plant operators.

Also, the number of operators required per shift, the number of shifts, the college education required for operators, the training in simulators, and emergency procedure revisions were all important improvements. So costs went up, reliability went down.

Q: Why was that?

A: Because plants were installing and backfitting new equipment, preparing new procedures, and retraining staff under tight deadlines. At the same time, industry realized that they could not tolerate another accident in the U.S., or public opinion would totally turn negative. Utilities formed their own organization to try to work together. Before TMI the utilities tended to work independently and didn't really share data much with each other. Afterward, they formed the Institute of Nuclear Power Operations, which was focused on improving performance of operations and the sharing of information about all equipment failures and operational occurrences. That organization has played a major role in improving the performance of all plants. Independent of the NRC, they began to improve each other, and developed better ways to run their plants.

So that took a while, but by the mid-1990s plants began to run more reliably. Their improved performance has resulted in the increased production of over 23,000 MW with no increase in installed nuclear generating capacity since the early 1990s.

Q: Do you think that the restructuring of the electricity markets that was started in the late 1990s has made it less likely that utilities will order new nuclear power plants because they can't pass the capital costs onto the consumers?

A: I thought that the financial restructuring might bring cost-cutting that would lessen safety, but it hasn't diminished safety. Utilities realize that safety and reliability are intertwined in that as plants operate more and more reliably, there are less and less challenges to the safety systems.

Q: But because they can't pass so-called stranded costs, the power producers tend to build natural gas plants. They like the natural gas plants because there are fewer up-front costs to building them. Exactly the opposite is true of nuclear power plants.

A: Stranded costs don't seem to be much of an issue today. When I left the NRC, I thought that natural gas plants were going to be the wave of the future for the power industry. At the time, one dollar would buy a million BTUs at the wellhead. The natural gas was cheap, widely available, and plants could be built in small, incremental steps. Also, it was easier for the utilities to finance 50 megawatts or 100 megawatts than a large coal or nuclear plant. Gas turbines could be built in a factory, shipped to the site, placed on a concrete pad and be operating in a relatively short period of time. From an environmental standpoint—this was before worries about CO_2—it was considered the cleanest fuel. Natural gas would likely still be the choice of utilities today for those reasons if prices had stayed low. But, with increased usage of natural gas, prices have gone up.

The price has recently been as high as $10 per million BTUs, so it's not as attractive.

The only two fuels that the U.S. has in abundance for future energy sources, in my view, are coal and nuclear. Coal faces a set of likely challenges in reducing carbon dioxide emissions. Nuclear power has not solved its problems with high-level waste storage. I think if you really want to make progress on global warming you've got to consider nuclear a prime candidate along with renewable energy sources.

When I was in school, I thought that practical fusion power was perhaps 20 years away. Now, 50 years later, it still seems at least 20 years away. It may make a breakthrough sometime, but it's been slow coming. It seems inevitable that the costs of natural gas are going to increase, and we're relying more and more on imported oil. We import more oil now than we did during the Carter era. One should not look at nuclear power in isolation but compare it to other alternatives.

Q: I notice that the NRC has four new designs—the AP1000, EPR, ESBWR, and the ABWR—that have been approved or are up for design approval and certification. All of these incorporate passive safety designs to a much larger degree than the reactors now in operation. How much safer do you think these new designs are?

A: After all the TMI improvements were incorporated into existing plants, they became, in my opinion, at least one or two orders of magnitude safer. It's hard to quantify because each plant is different, as is the value of the backfitting that was done. Considerable attention is now given to proper operation and to preventing and mitigating severe accidents. Plants are running much safer today, as indicated, for example, by reduced equipment failures, operator errors, and safety systems activations.

But [after TMI] reactor designers went back to the drawing board. These new advanced designs rely more on passive systems for emergencies, are simpler, have increased margins [of safety], and are safer. Some designs have now been approved by the Commission as standard plants. These are another level up in safety from today's plants. Certainly they have features which increase the level of safety and take into account all the worldwide experience with power plants that wasn't available when the earlier generation of plants were built.

Q: Listening to you and knowing that there is so much more talk about building nuclear power plants now than there was even five years ago, when do you suppose the next nuclear power reactor will be ordered?

A: It seems a little easier to predict the location of the next nuclear plant than to predict when one might be built. The most interested utilities tend to be in southern states. These states have increasing populations and increasing needs for electrical power. An arc of coastal states from Texas to Virginia seems to describe where interest in nuclear is highest. What would tilt the entire energy picture, of course, would be adoption by Congress of a carbon emission tax. Coal plants, oil plants, and natural gas plants would face additional costs.

I would like to mention a few other things about nuclear power. Years after TMI, when the utility was able to get access to the reactor core, they found that half the core had melted the day of the accident. I had assumed that if a significant amount of fuel melted it would probably penetrate the reactor vessel and spill molten material on the floor of the containment building. Later examinations showed that half the core did melt and the half-molten core did not penetrate the reactor vessel. Also, after TMI the Commission's research division instituted large-scale programs to assess severe accidents and containment performance. Experiments included pouring simulated molten cores onto simulated containment floors and tests to determine the actual strength of containments. A conclusion was that the TMI containment vessel would not have lost its integrity even if there had been complete core melting. The molten fuel would have spread, been diluted by the materials in the floor and solidified in place without causing containment failure. Thus, the much-discussed China Syndrome would not have occurred.

Epidemiological studies by the University of Pittsburgh have found no evidence of any health effects from the TMI accident.

The new designs go even further to try to mitigate severe accidents and lengthen the amount of time available before a core overheats. They have larger safety margins and more passive systems. In my opinion the present generation of plants is safer than ever. They demonstrate that in their day-to-day operation, and the next generation of reactors, if there are any built, will be even safer.

Q: Some are being built now in Japan.

A: Yes, and in France. Another one of the most modern plants in the world is being built in Finland, which is using a design by AREVA and Siemens. The most modern research reactor in the world is being built in Australia. Modern plants will be built in China too. So to some extent while we weren't building, other countries went ahead and kept building, and the most advanced designs today, and the most operating experience with these new designs, will come from foreign plants.

Q: Yes. The pressurized water and boiling water reactors were initially U.S. designs.

A: Yes, an infrastructure once existed in the United States to build nuclear power plants. Many of the components for new designs will come from those countries that are continuing to rely on nuclear power, such as Japan and France.

From my standpoint nuclear plants can be made safer. They will be made safer on into the future, but even now they are quite safe compared with almost any other energy source. If global warming will be occurring over the next 50 to 100 years, there is no way to significantly reduce our contribution to carbon emissions other than by the use of nuclear plants. The amount of energy released per chemical reaction is quite small compared to the amount of energy that is released per fission reaction. The maximum chemical reaction releases between one and ten electron volts per atom. Uranium fission releases about 200 million electron volts. Thus, about 100 million times more energy is available on a per-atom basis from uranium than from fossil fuels. Nuclear power remains controversial, but utilities are showing interest again in nuclear. In my opinion, nuclear power is a source that is available now for the safe generation of electricity.

Q: Mr. Denton, thank you very much for sharing your time and your insights.

Appendix

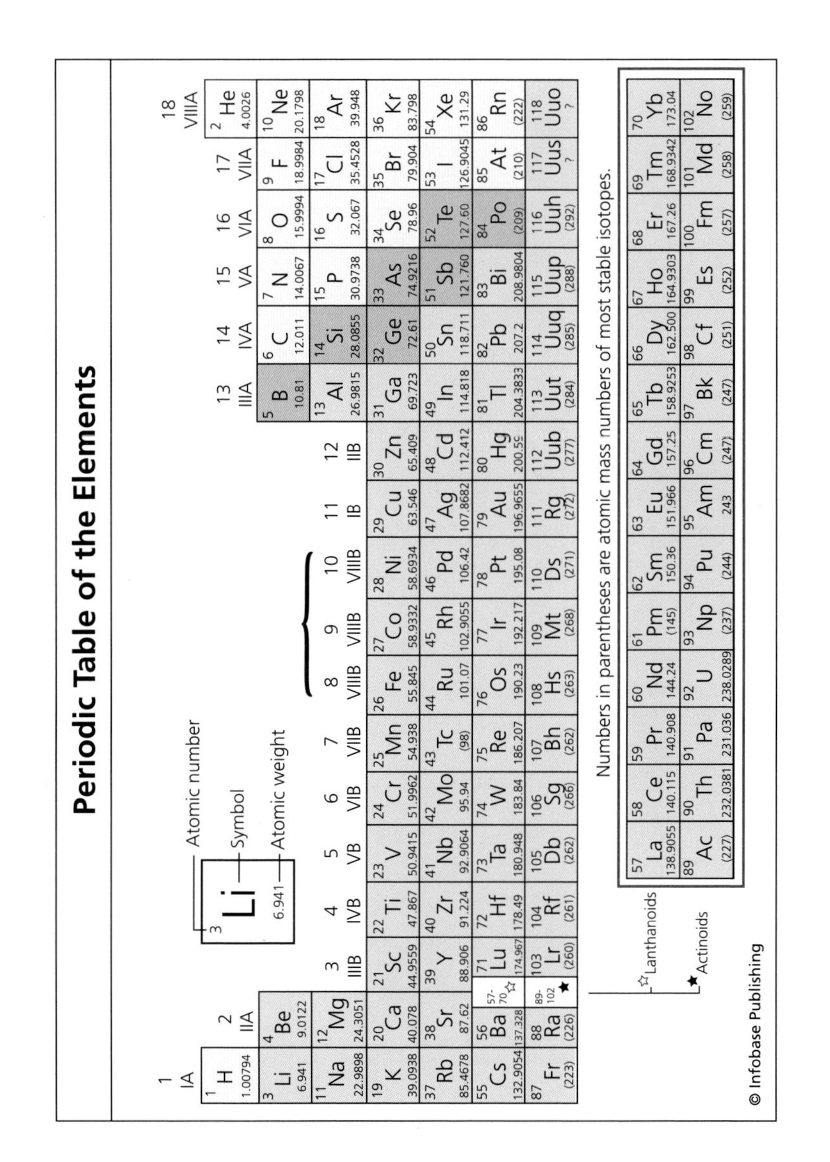

Periodic Table of the Elements

Key: Atomic number (3), Symbol (Li), Atomic weight (6.941)

1 IA	2 IIA	3 IIIB	4 IVB	5 VB	6 VIB	7 VIIB	8 VIIIB	9 VIIIB	10 VIIIB	11 IB	12 IIB	13 IIIA	14 IVA	15 VA	16 VIA	17 VIIA	18 VIIIA
1 H 1.00794																	2 He 4.0026
3 Li 6.941	4 Be 9.0122											5 B 10.81	6 C 12.011	7 N 14.0067	8 O 15.9994	9 F 18.9984	10 Ne 20.1798
11 Na 22.9898	12 Mg 24.3051											13 Al 26.9815	14 Si 28.0855	15 P 30.9738	16 S 32.067	17 Cl 35.4528	18 Ar 39.948
19 K 39.0938	20 Ca 40.078	21 Sc 44.9559	22 Ti 47.867	23 V 50.9415	24 Cr 51.9962	25 Mn 54.938	26 Fe 55.845	27 Co 58.9332	28 Ni 58.6934	29 Cu 63.546	30 Zn 65.409	31 Ga 69.723	32 Ge 72.61	33 As 74.9216	34 Se 78.96	35 Br 79.904	36 Kr 83.798
37 Rb 85.4678	38 Sr 87.62	39 Y 88.906	40 Zr 91.224	41 Nb 92.9064	42 Mo 95.94	43 Tc (98)	44 Ru 101.07	45 Rh 102.9055	46 Pd 106.42	47 Ag 107.8682	48 Cd 112.412	49 In 114.818	50 Sn 118.711	51 Sb 121.760	52 Te 127.60	53 I 126.9045	54 Xe 131.29
55 Cs 132.9054	56 Ba 137.328	57-70 ☆	72 Hf 178.49	73 Ta 180.948	74 W 183.84	75 Re 186.207	76 Os 190.23	77 Ir 192.217	78 Pt 195.08	79 Au 196.9655	80 Hg 200.59	81 Tl 204.3833	82 Pb 207.2	83 Bi 208.9804	84 Po (209)	85 At (210)	86 Rn (222)
87 Fr (223)	88 Ra (226)	89-102 ★	104 Rf (261)	105 Db (262)	106 Sg (266)	107 Bh (262)	108 Hs (263)	109 Mt (268)	110 Ds (271)	111 Rg (272)	112 Uub (277)	113 Uut (284)	114 Uuq (285)	115 Uup (288)	116 Uuh (292)	117 Uus ?	118 Uuo ?

Group 3 also contains 71 Lu 174.967 (period 6) and 103 Lr (260) (period 7).

☆ Lanthanoids

57 La 138.9055	58 Ce 140.115	59 Pr 140.908	60 Nd 144.24	61 Pm (145)	62 Sm 150.36	63 Eu 151.966	64 Gd 157.25	65 Tb 158.9253	66 Dy 162.500	67 Ho 164.9303	68 Er 167.26	69 Tm 168.9342	70 Yb 173.04

★ Actinoids

89 Ac (227)	90 Th 232.0381	91 Pa 231.036	92 U 238.0289	93 Np (237)	94 Pu (244)	95 Am 243	96 Cm (247)	97 Bk (247)	98 Cf (251)	99 Es (252)	100 Fm (257)	101 Md (258)	102 No (259)

Numbers in parentheses are atomic mass numbers of most stable isotopes.

The Chemical Elements

(g) none (c) nonmetallics

element	symbol	a.n.
carbon	C	6
hydrogen	H	1

(g) chalcogen (c) nonmetallics

element	symbol	a.n.
oxygen	O	8
polonium	Po	84
selenium	Se	34
sulfur	S	16
tellurium	Te	52
ununhexium	Uuh	116

(g) alkali metal (c) metallics

element	symbol	a.n.
cesium	Cs	55
francium	Fr	87
lithium	Li	3
potassium	K	19
rubidium	Rb	37
sodium	Na	11

(g) alkaline earth metal (c) metallics

element	symbol	a.n.
barium	Ba	56
beryllium	Be	4
calcium	Ca	20
magnesium	Mg	12
radium	Ra	88
strontium	Sr	38

(g) none (c) metallics

element	symbol	a.n.	element	symbol	a.n.
aluminum	Al	13	scandium	Sc	21
bohrium	Bh	107	seaborgium	Sg	106
cadmium	Cd	48	silver	Ag***	47
chromium	Cr	24	tantalum	Ta	73
cobalt	Co	27	technetium	Tc	43
copper	Cu***	29	thallium	Tl	81
darmstadtium	Ds	110	titanium	Ti	22
dubnium	Db	105	tin	Sn	50
gallium	Ga	31	tungsten	W	74
gold	Au***	79	ununbium	Uub	112
hafnium	Hf	72	ununtrium	Uut	113
hassium	Hs	108	ununquadium	Uuq	114
indium	In	49	vanadium	V	23
iridium	Ir ****	77	yttrium	Y	39
iron	Fe	26	zinc	Zn	30
lawrencium	Lr	103	zirconium	Zr	40
lead	Pb	82			
lutetium	Lu	71			
manganese	Mn	25			
meitnerium	Mt	109			
mercury	Hg	80			
molybdenum	Mo	42			
nickel	Ni	28			
niobium	Nb	41			
osmium	Os****	76			
palladium	Pd****	46			
platinum	Pt ****	78			
rhenium	Re	75			
rodium	Rh****	45			
roentgenium	Rg	111			
ruthenium	Ru****	44			
rutherfordium	Rf	104			

(g) pnictogen (c) metallics

element	symbol	a.n.
arsenic	As*	33
antimony	Sb*	51
bismuth	Bi	83
nitrogen	N	7
phosophorus	P**	15
ununpentium	Uup	115

(g) none (c) semimetallics

element	symbol	a.n.
boron	B	5
germanium	Ge	32
silicon	Si	14

(g) actinoid (c) metallics

element	symbol	a.n.
actinium	Ac	89
americium	Am	95
berkelium	Bk	97
californium	Cf	98
curium	Cm	96
einsteinium	Es	99
fermium	Fm	100
mendelevium	Md	101
neptunium	Np	93
nobelium	No	102
plutonium	Pu	94
protactinium	Pa	91
thorium	Th	90
uranium	U	92

(g) lanthanoid (c) metallics

element	symbol	a.n.
cerium	Ce	58
dysprosium	Dy	66
erbium	Er	68
europium	Eu	63
gadolinium	Gd	64
holmium	Ho	67
lanthanum	La	57
neodymium	Nd	60
praseodymium	Pr	59
promethium	Pm	61
samarium	Sm	62
terbium	Tb	65
thulium	Tm	69
ytterbium	Yb	70

(g) halogens (c) nonmetallics

element	symbol	a.n.
astatine	At*	85
bromine	Br	35
chlorine	Cl	17
fluorine	F	9
iodine	I	53
ununseptium	Uus*	117

(g) noble gases (c) nonmetallics

element	symbol	a.n.
argon	Ar	18
helium	He	2
krypton	Kr	36
neon	Ne	10
radon	Rn	86
xenon	Xe	54
ununoctium	Uuo	118

a.n. = atomic number
(g) = group
(c) = classification

* = semimetallics (c)
** = nonmetallics (c)
*** = coinage metal (g)
**** = precious metal (g)

Chronology

1896 Antoine-Henri Becquerel's experiments with uranium leads him to discover radioactivity.

1897 Marie and Pierre Curie discover the radioactive elements polonium and radium.

1902 Ernest Rutherford and Frederick Soddy discover that radioactivity is a process by which atoms of one type of element spontaneously disintegrate to produce atoms of other elements.

1905 Albert Einstein publishes his first theory of relativity, a consequence of which is the equation $E = mc^2$, which explains why the fission of a nucleus releases such large amounts of energy.

1913 H. G. Wells publishes his novel *The World Set Free: A Story of Mankind* in which he describes the use of atomic energy in war and peace.

Niels Bohr first proposes the model of the atom that bears his name. His ideas are radically different from those of most of his contemporaries.

1934 Enrico Fermi bombards uranium with neutrons to create the first artificial element, neptunium.

1938 Otto Hahn and Fritz Strassmann bombard uranium with neutrons and produce barium and krypton, an experiment that leads to the discovery of atomic fission.

1941 Glenn T. Seaborg, Joseph W. Kennedy, and Arthur C. Wahl, at University of California, Berkeley, create plutonium for the first time.

1942 President Franklin D. Roosevelt gives approval for the Manhattan Project, the wartime effort to manufacture the first atomic bomb.

The world's first self-sustaining nuclear reaction is achieved at the University of Chicago under the direction of Enrico Fermi.

1945 The first atomic bomb is detonated at Alamogordo, New Mexico. It is fueled by plutonium. This is followed by the destruction of Hiroshima, Japan, by a uranium-fueled bomb, and a few days later by the destruction of Nagasaki, Japan, by a plutonium-fueled bomb.

1946 U.S. Atomic Energy Commission, the forerunner to the Nuclear Regulatory Commission, is established.

1951 Experimental Breeder Reactor 1, the first nuclear reactor to generate electricity, begins operation in Arco, Idaho. It is now a National Historic Landmark.

1953 President Dwight D. Eisenhower delivers his famous "Atoms for Peace" speech in which he calls for the development and the sharing of peaceful nuclear technology.

1954 The Soviet Union starts the Obninsk Nuclear Power Plant, the world's first nuclear reactor used to produce useful amounts of electricity. It has an output of 5 MWe.

1955 Arco, Idaho, population 1,000, becomes the first town to be lit entirely by a nuclear reactor. The experimental BORAX III, a boiling water reactor, was used to generate the power.

1956 The first commercial nuclear reactor, located at Calder Hall, England, goes online.

1957 The Shippingport, Pennsylvania, nuclear reactor, the first commercial nuclear power plant in the United States, goes online.

1962 First CANDU reactor goes online.

1966 The German AVR, the first pebble bed nuclear reactor, goes online.

1973 The French Phénix, a fast-neutron, sodium-cooled, large-scale breeder reactor, goes online.

1974 The Atomic Energy Commission is reorganized into the Nuclear Regulatory Commission and the now-defunct Energy Research and Development Administration.

1977 President Carter announces that the United States will not pursue the reprocessing of spent nuclear fuel.

1979 The accident at Three Mile Island occurs.

1982 The Shippingport nuclear reactor is permanently shut down.

1983 President Reagan signs the Nuclear Waste Policy Act of 1982, and the federal government begins a research effort that is supposed to result in the creation of a national high-level storage site for radioactive waste.

1986 The Chernobyl nuclear power plant accident occurs.

1988 The AVR pebble bed nuclear reactor is permanently closed.

1991 Japan begins construction of the first of several Advanced Boiling Water Reactors, a so-called Generation III design.

1993 China's first commercial nuclear reactor, Qinshan 1, is connected to the grid.

1996 Taiwan orders two Advanced Boiling Water Reactors.

2000 The South African company Eskom is established to build and market pebble bed modular reactors.

2002 Finland begins construction of a Generation III pressurized water reactor.

2004 China's *Peoples Daily* reports plans to bring two reactors online per year for the next 16 years.

2007 The U.S. Nuclear Regulatory Commission announces that it expects applications for 12 new reactors during 2007 and applications for an additional 15 reactors during 2008.

List of Acronyms

ABWR Advanced Boiling Water Reactor

ALARA as low as reasonably achievable

BWR boiling water reactor

CANDU name for pressurized heavy water reactor pioneered in Canada

CEA Commissariate à l'énergie atomique

DOE U.S. Department of Energy

EPA U.S. Environmental Protection Agency

IAEA International Atomic Energy Agency

ISO independent system operator

LWR light water reactor

MOX mixed oxide fuel

NPT Treaty on the Non-Proliferation of Nuclear Weapons

NRC Nuclear Regulatory Commission

OPEC Organization of Petroleum Exporting Countries

PBMR pebble bed modular reactor

PRA probabilistic risk assessment

PWR pressurized water reactor

RERF Radiation Effects Research Foundation

TMI Three Mile Island (nuclear reactor accident)

activity the intensity of radioactive decay

alpha particle the nucleus of a helium atom consisting of two protons and two neutrons. It is ejected at high speed from the atomic nucleus.

atomic number the number of protons in the nucleus of an atom

beta particle (also called a beta ray) an electron ejected at high speed from the nucleus of an atom during certain transmutation events

breeder reactor nuclear reactors that produce more reactor fuel than they consume

calandria the large vessel that contains the fuel, coolant, and moderator in a CANDU reactor

capacity factor the percentage of time that, on average, a power plant is operated at full power

capital costs the total costs incurred during the process of bringing a project into commercial operation

centrifuge a device used to separate materials of different densities

conservation law a physical principle, or law of nature, that asserts that a particular property, (e.g., energy) remains constant in any isolated system

control rod a rod used to regulate reactor output by absorbing neutrons that might otherwise cause additional fissions

criticality accident any accident that is caused by the loss of control of the fission rate

dose-response curve any curve that displays the "response," usually interpreted as a cancer rate, as a function of the "dose," a measure of radiation exposure

efficiency (also thermal efficiency) a measure of the useful output of a heat engine, (e.g., electricity), to the amount of thermal energy required to produce that output

electron a subatomic particle carrying a single negative electrical charge

energy density the amount of energy produced per unit mass of fuel

enriched uranium a sample of uranium that has been processed so that the proportion of uranium-235 to uranium-238 is greater than that found in nature

fast breeder reactor a breeder reactor that uses fast neutrons (as opposed to thermal neutrons)

fast neutron a neutron that is not in thermal equilibrium with its surroundings

fertile an isotope whose atoms can be readily converted into fissile material

fissile an isotope whose atoms can readily undergo fission

fission the splitting or division of an atomic nucleus resulting in the release of large amounts of energy

fuel assembly a structure designed to hold fuel rods in such a way as to facilitate (1) the insertion and removal of fuel into the reactor, (2) the circulation of coolant around the fuel rods, (3) and the circulation of the moderator if a moderator is required

fuel rod a hollow tube designed to (1) contain the reactor fuel, (2) contain the products of fission, and (3) facilitate the transfer of heat from the fuel to the surrounding medium

half-life the time required for half the atoms in a sample of radioactive material to disintegrate. The half-life of a substance is independent of the size of the sample.

heat engine a device to convert thermal energy into work or electrical energy

heat exchanger a device that permits the transfer of heat but not mass between two fluids

isotope any collection of atoms all of which have identical atomic numbers and identical mass numbers

kinetic energy the energy associated with motion

linear hypothesis the assertion that the dose-response curve is a straight line that begins at the origin of the dose-response coordinate system and remains straight for very high doses of radiation

loss-of-coolant accident a reactor accident in which the ability of the system to cool itself is compromised

mass number the arithmetic sum of all protons and neutrons in a nucleus

moderator any material the purpose of which is to convert fast neutrons to thermal neutrons

negative void coefficient a characteristic of certain types of reactors that causes the fission rate to decrease as the rate at which thermal energy is produced increases

neutron a subatomic particle usually found in the nuclei of atoms, having approximately the same mass as a proton, and containing zero net electrical charge

nucleus the positively charged center of an atom consisting of a tightly packed mass of protons and (usually) neutrons

once-through fuel cycle the policy that requires that nuclear fuel be used only once. In such a cycle, the fuel is not reprocessed.

passive safety those reactor control mechanisms that operate without operator intervention and that depend solely on the physical characteristics of the materials from which the reactor is constructed

pressure vessel the structure that contains the reactor fuel and coolant, usually at elevated pressures

proton a positively charged subatomic particle found in the nuclei of all atoms

radioactive decay the process by which certain nuclei disintegrate while spontaneously emiting alpha particles, beta particles, and/or gamma rays

radioactivity the property of undergoing radioactive decay

reprocessing the process by which uranium and plutonium are recovered from spent reactor fuel to be used again as fuel

sievert a unit of measurement of radiation designed to take into account the biological effects produced by the type of radiation absorbed

spot market the system by which power is purchased by electric utilities under short-term contracts

steam generator a device used in pressurized water reactors to transfer heat but not mass between two supplies of water for the purpose of inducing a phase change in the cooler water from liquid to steam

thermal energy the average energy of particles in a system at uniform temperature

thermal neutron a neutron whose average kinetic energy equals that of the other particles in the surrounding medium

transmutation the conversion of one element into another without the process of fission

turbine a device for converting the kinetic energy of a moving fluid into rotary motion

Further Resources

Nuclear power is a rich subject that can be studied from a variety of viewpoints. Below is a sampling of some of the many ways that this topic can be explored.

BOOKS

Bodansky, David. *Nuclear Energy: Principles, Practices, and Prospects.* 2d ed. New York: Springer, 2004. The author wrote this book for people "with technical backgrounds and those without such backgrounds." Readers of the present volume are well prepared to tackle this book's more than 600 pages. It is the most comprehensive book of its kind accessible to the general reader.

Henderson, Harry. *Nuclear Power: A Reference Handbook.* Santa Barbara, Calif.: ABC-CLIO, 2000. An excellent source book for things nuclear.

Josephson, Paul R. *Red Atom: Russia's Nuclear Power Program from Stalin to Today.* New York: W.H. Freeman, 2000. The Soviet Union produced a long line of brilliant engineers and scientists as well as a terrible culture of nuclear safety. This fascinating book describes both.

Kirkland, Kyle. *Particles and the Universe.* New York: Facts On File, 2007. More information on nuclear physics, the branch of science that is, in part, concerned with the reactions that power nuclear reactors.

National Research Council. *Safety and Security of Commercial Spent Nuclear Fuel Storage: Public Report.* Washington, D.C.: National Academies Press, 2006. A brief highly readable report on a very important topic.

National Research Council. *Technical Bases for Yucca Mountain Standards.* Washington, D.C.: National Academies Press, 1995. A careful discussion of the possibility of accurately predicting repository performance far into the future.

Nuclear Energy Agency, Organisation for Economic Co-operation and Development. *Nuclear Energy Today.* Paris: OECD Publications, 2003. An excellent and easily accessible introduction to the subject.

Pool, Robert. *Beyond Engineering: How Society Shapes Technology.* New York: Oxford University Press, 1997. An interesting book that is concerned with the general problem of how society affects technology. This is a subject with particular relevance to the development of nuclear power.

Sweet, William. *Kicking the Carbon Habit: Global Warming and the Case for Renewable and Nuclear Energy.* New York: Columbia University Press, 2006. This book's primary focus is on reducing carbon emissions. It examines nuclear energy in this context.

INTERNET RESOURCES

Some of the following Web sites contain hundreds or even thousands of pages of information:

Allardice, Corbin and Edward R. Trapnell. "The First Pile." Available online. URL: www.osti.gov/accomplishments/pdf/DE00782931/DE00782931.pdf. Accessed on August 18, 2008. A dramatic account of the world's first atomic reactor.

CANDU Owners Group. "CANDU Reactors." Available online. URL: http://www.candu.org/candu_reactors.html. Accessed on April 15, 2007. This site provides a good overview of the history and technology of CANDU reactors.

International Atomic Energy Agency. Available online. URL: http://www.iaea.org/. Accessed on April 15, 2007. An excellent resource for learning about specific applications of nuclear energy and for obtaining an overview of nuclear power around the world. Especially interesting is their report on Chernobyl: URL: http://www.iaea.org/Publications/Booklets/Chernobyl/chernobyl.pdf. Accessed on April 15, 2007.

U.S. Department of Energy. *GenIV Nuclear Energy Systems.* Available online. URL: http://nuclear.energy.gov/genIV/neGenIV1.html. Accessed on April 2, 2007. The site contains the Department of Energy's description of Generation IV reactors. The result of international collaboration, some of these designs may be the next big thing in electrical energy production.

U.S. Nuclear Regulatory Commission. Available online. URL: http://www.nrc.gov/reading-rm/basic-ref.html. Accessed on April 15, 2007. The Nuclear Regulatory Commission maintains an enormous collection of documents about all aspects of nuclear energy. Some were written for children, some for the general reader, and some for nuclear engineers. Its reference librarians, who answer inquiries promptly, are also very helpful.

VIDEOS

Cameco Corporation. *Mining and Underground Processing of Cigar Lake Ore.* Available online. URL: http://www.cameco.com/operations/uranium/cigar_lake/mining_and_milling.php. Accessed on April 14, 2008. This fascinating four-minute video provides an overview of the method by which uranium ore is extracted at the Cigar Lake uranium mine, site of one of the world's richest deposits of ore.

U.S. Department of Energy. *Splitting Atoms: An Electrifying Experience.* Available online. URL: http://www.ne.doe.gov/publicInformation/nePublicInformation2a.html#videos. Accessed on April 15, 2007. This 11-minute film, aimed at a younger audience, is an entertaining and informative overview of nuclear power. (Requires a fast Internet connection and RealPlayer.)

———. *A Vision for Nuclear-Enabled Peace and Prosperity.* Available online. URL: http://www.ne.doe.gov/publicInformation/nePublicInformation2a.html#videos. Accessed on April 3, 2007. An interesting description of future U.S. plans in the area of nuclear energy.

Index

Note: *Italic* page numbers indicate illustrations and diagrams; page numbers followed by *c* indicate chronology entries; page numbers followed by *t* indicate charts, graphs, or tables.

A

O

T

U